THE DATA DIVIDEND
YOUR CLICKS FUEL AI. WHY AREN'T YOU GETTING PAID?

EUSTACE ASANGHANWA

Copyright © 2025 by Visionary Quill, LLC

All rights reserved.

No part of this book may be reproduced in any form or by any electronic or mechanical means, including information storage and retrieval systems, without written permission from the author, except for the use of brief quotations in a book review.

In loving memory of
WrayAdy Shu Asanghanwa
Forever missed, never forgotten,
And always a source of strength and inspiration.

INTRODUCTION TO THE DATA DIVIDEND

Every major technological revolution has relied on a labor force. The feudal age had serfs tilling land for lords. The industrial age had miners, machinists, and assembly line workers. The digital age has us—clicking, scrolling, uploading, correcting. But this labor force is invisible. Our behaviors fuel predictive engines and train artificial intelligence (AI) systems, yet few realize they're working at all.

This book is divided into three parts:

PART I: THE GREAT DATA HEIST (CHAPTERS 1–4)

We expose how everyday behavior has quietly become the backbone of AI—and how this unpaid labor is already fueling economic and job upheaval worldwide. Clicks, likes, and swipes are not harmless digital traces. They are the raw materials of trillion-dollar industries.

PART II: THE TECHNOLOGY OF ECONOMIC JUSTICE (CHAPTERS 5–8)

If Part I revealed the problem, this part presents a blueprint for change. We introduce technologies that can rebalance the digital economy—tools like smart contracts, personal data vaults, and confidential computing that let users retain control, enforce consent, and share in value creation. It is about building infrastructure that makes data dignity real.

PART III: BEYOND UNIVERSAL BASIC INCOME (CHAPTERS 9–11)

Money alone won't restore agency. UBI may ease the pain of displacement, but only participatory models restore power to people. This final section argues for dignity over dependency and makes the strategic case for why tech companies—and societies—must embrace the data dividend. The book concludes with actions readers can take now to begin shifting the digital economy.

Together, these three parts will guide you from recognition of an invisible injustice to a vision for reclaiming what is rightfully yours: a fair share of the digital wealth you help create.

PART ONE
THE GREAT DATA HEIST

CHAPTER 1
THE INVISIBLE WORKFORCE

MARIA'S DISPLACEMENT

MARIA ALVAREZ WAKES BEFORE DAWN, the sound of her alarm cutting through the silence of her small apartment in Las Vegas. For 18 years, she worked as a housekeeper at one of the city's many hotels. Her days were filled with quiet labor—making beds, scrubbing bathrooms, restocking soaps. Guests rarely saw her. When they did, she smiled politely and moved on.

In late 2024, her hotel announced it was adopting a new suite of AI-powered cleaning robots. The general manager assured staff this technology would "assist, not replace" them. Within three months, half the housekeeping staff—Maria included—were laid off.

What stung the most wasn't just losing her job, but realizing that years of her movements, habits, and even route timings through hotel corridors had helped train the very AI that made her redundant. Her badge swipe data, room-cleaning logs, wearable productivity trackers—all had quietly become part of a corporate dataset.

Maria was invisible twice over: first as a worker, then as a contributor to a data system that erased her.

She's not alone. Around the world, billions of people generate training data for artificial intelligence just by living online. This unpaid, unrecognized digital labor has powered one of the most transformative technological revolutions of our time—and yet the workers remain invisible.

You are not the product. You are the source.

∼

YOUR CLICKS ARE WORK

Artificial intelligence doesn't emerge from a vacuum. It learns from patterns—patterns derived from human behavior. Every time you type a query into a search engine, correct a voice assistant, or scroll past a post on social media, you're contributing to a training set. You are, in effect, working.

This labor is not optional. Participation in the modern economy requires it. Signing up for a service, making a purchase, or simply existing online creates behavioral residue—what companies call "engagement" and algorithms treat as fuel.

But unlike traditional labor, this kind of work is uncompensated. There are no wages, no contracts, no acknowledgments. Most people don't even know they're working.

THE NEW OIL RUSH

In the early 2000s, data was famously dubbed "the new oil." But unlike oil, which is extracted from the earth, this resource is extracted from people. And unlike oil workers, the producers of this resource are not paid.

Consider social media: platforms like Facebook, Instagram, and TikTok generate immense value from user activity. Your posts, comments, likes, and even the time you linger on a video are all signals that fine-tune algorithms. These algorithms power targeted ads, which generate billions in revenue.

You see entertainment. They see labor.

THE RISE OF THE BEHAVIORAL ECONOMY

What began as tracking for improvement has become a default business model. Companies no longer just collect data—they analyze, trade, and monetize it at scale. AI models, especially large language models like GPT and image generators like Midjourney, are trained on oceans of human-generated content: Reddit threads, product reviews, Wikipedia edits, forum questions, and more.

These contributions are scraped, sampled, and fed into systems that now perform tasks once reserved for humans—writing copy, composing music, generating art, and even coding.

The people who produced the data? Often unaware. Always unpaid.

THIS IS LABOR

It may not feel like work. There's no boss, no office, no clock-in. But that's part of what makes it extractive. Digital labor happens ambiently, invisibly. And it happens *at scale*—across demographics, borders, and time zones.

A rideshare driver teaches the route optimizer.
A TikTok dancer trains the recommendation engine.
A gamer refines the toxicity filter.
A shopper powers the predictive inventory model.
The value created is real. But the labor goes unrecognized.

THE DISPLACEMENT CYCLE

Here's the paradox: the more data we produce, the smarter AI systems become. And the smarter they become, the more they can automate away the very jobs that people rely on.

Teachers, coders, designers, paralegals, and customer service reps are already seeing tasks shifted to AI. Yet many of those same systems were trained, at least in part, on the behavioral footprints of those very workers.

We are building the tools that will replace us—without agency, consent, or compensation.

NAMING WHAT'S BEEN HIDDEN

To fix this, we must first name it: **data is labor**.

That recognition changes everything. It repositions the public not as passive users but as active contributors. It reframes the tech economy not as magical and meritocratic but as extractive and imbalanced.

And it opens the door to something powerful: a demand for fairness.

This book is about reclaiming dignity in the age of AI. It's about making the invisible workforce visible—and ultimately, compensated. Because if our clicks are building machines, then we deserve more than targeted ads and job loss in return.

We deserve a **data dividend**.

And we're just getting started.

KEY TAKEAWAYS

- Billions of people are unknowingly part of the AI supply chain.

- Everyday actions like searches, clicks, and photos are training data for AI.
- This digital labor is unpaid, yet it fuels a trillion-dollar industry.
- Job displacement often follows the data trail—the trainers are the first to be replaced.
- Without recognition and rights, we are digital sharecroppers in the AI economy.

LOOKING AHEAD

In the next chapter, we'll explore how platforms like Stack Overflow, GitHub, and social media forums—once thriving hubs of volunteer knowledge-sharing—are being hollowed out by AI systems built on that very content. This isn't just about job loss. It's about value extraction without reinvestment—and the death of the digital commons.

CHAPTER 2

YOU BUILT THE MACHINES THAT REPLACED YOU

THE STACK OVERFLOW PARADOX

FOR OVER A DECADE, Stack Overflow was the sacred archive of programming knowledge. Millions of developers contributed code snippets, bug fixes, and best practices—not for money, but for community, recognition, and the joy of helping others.

Then came a new wave of AI-powered coding assistants.

Launched in 2021, tools like GitHub Copilot and ChatGPT began offering real-time code generation based on natural language prompts. These tools were trained on publicly available code, including open-source repositories and volumes of programming Q&A content from platforms like Stack Overflow.

By 2023, many developers recognized a paradox: AI assistants were built on their unpaid labor—yet these very tools were reducing visits to the communities they came from. Stack Overflow, facing declining engagement, restricted unauthorized data scraping and launched a licensing program. But without fresh contributions, the knowledge base began to stagnate.

It's a cycle of extraction without reinvestment. A digital commons drained dry.

When knowledge becomes legacy data, labor becomes legacy too.

∾

DIGITAL INTERACTIONS BECOME TRAINING DATA

Every click, correction, and content post becomes raw material for machine learning. Users aren't just interacting with products—they're producing the intelligence that powers them.

And yet, most of this happens without consent, transparency, or compensation.

AI LEARNS BY WATCHING YOU WORK

To generate content, solve problems, or mimic creativity, AI needs examples. It learns by watching us work—on forums, in customer support chats, in videos and comments.

The teachers of AI are the very people it is beginning to replace.

WHEN LABOR BECOMES LEGACY DATA

The irony is sharp: the better you are at your job, the more valuable your data. The more efficient your code, the more persuasive your copy, the more insightful your strategy—the more likely that your work will be scraped and reused without attribution.

Welcome to the age of legacy labor: where excellence becomes a blueprint for automation.

CALL CENTERS AND VOICE ASSISTANTS: A CASE IN POINT

Tens of thousands of real conversations between agents and customers have trained the most advanced AI voice systems. As these systems improve, they displace the very workers who made them smart.

The annotations, rebuttals, and tone modulations of entry-level agents become competitive advantages—for machines, not humans.

NOT JUST BLUE COLLAR

AI automation doesn't just threaten manual labor. It's increasingly replacing knowledge workers: journalists, designers, marketers, translators. Their past output—articles, campaigns, translations—now trains the tools that edge them out.

Their labor doesn't just influence automation. It enables it.

A SYSTEM WITHOUT FEEDBACK LOOPS

In traditional economies, labor creates value—and value flows back to the laborer. In the data economy, the loop is broken. Platforms harvest contributions but provide no credit, no compensation, no control.

This is enclosure, not empowerment.

KEY TAKEAWAYS

- AI systems are trained on user-generated content and behavior—often without consent.
- The more skilled or creative the user, the more valuable (and vulnerable) their contributions.

- Professions are being hollowed out by tools built from their own behavioral legacy.
- The current system rewards extraction without reinvestment or recognition.

LOOKING AHEAD

In the next chapter, we'll look at what happens when this cycle scales. From ride-hailing to retail, we'll follow the automation avalanche—and the overlooked role that human behavioral data plays in driving it.

CHAPTER 3
THE AUTOMATION AVALANCHE

THE VANISHING ROUTES

TANISHA IS a delivery driver in Minneapolis. For three years, she relied on a mix of DoorDash and Instacart gigs to make ends meet. She was efficient, polite, and highly rated. But in early 2025, her delivery volume dropped. At first, she thought it was a temporary glitch—then she realized she was being edged out by predictive dispatch algorithms.

The systems had learned from her behavior: which orders she accepted, what routes she preferred, how long deliveries took, when she took breaks. Now, new drivers were receiving the optimal routes she once dominated, and autonomous delivery robots were being tested on her former turf.

Tanisha had unknowingly trained the system that made her redundant. Her data became her own competition.

The smarter the system, the fewer the shifts.

FORECASTS THAT WERE ONCE THEORETICAL ARE NOW REAL

For years, automation forecasts were treated with a mix of skepticism and awe. Today, many of those projections are materializing:

- The **McKinsey Global Institute** projected that by 2030, up to **800 million jobs** could be displaced globally by automation [1].
- The **OECD** found that 14% of jobs across its member countries are highly automatable, with another 32% at risk of significant change [2].
- A **2023 Goldman Sachs report** estimated that AI alone could impact **300 million full-time jobs**, especially in administrative and legal sectors [3].

Industries at immediate risk include retail, transportation, customer service, hospitality, logistics, and creative work. Automation isn't just coming—it's accelerating.

THE DATA BEHIND THE DISPLACEMENT

What's often overlooked is the source of AI's capabilities: human behavioral data.

- **Instacart and DoorDash** use delivery driver behavior to optimize routes, reduce delivery times, and automate dispatch [4][5].
- **Amazon** tracks warehouse worker movements to inform robotics and workflow design [6].
- **Netflix** and **Spotify** refine recommendations through billions of micro-interactions with users [7][8].

These systems aren't born smart. They're trained—by people like Tanisha.

IT'S NOT JUST JOBS—IT'S BARGAINING POWER

Even when automation doesn't eliminate jobs, it often **weakens worker leverage**. Platforms use AI to:

- Predict demand and dynamically adjust pay.
- Enforce algorithmic scheduling.
- Monitor worker behavior to discourage collective action.

A 2021 study from **Data & Society** showed that ride-hailing apps used predictive analytics not only to assign rides, but also to influence when and how long drivers stayed online—managing a workforce without employing it [9].

This quiet managerial control erodes human agency.

A WIDER GAP: INEQUALITY IN THE WAKE OF AI

Automation amplifies inequality. It doesn't replace everyone equally:

- In the U.S., **Black and Latino workers** are overrepresented in automatable roles like cashiers and drivers.
- **Women** dominate clerical and caregiving jobs that are now being targeted by AI augmentation.

Meanwhile, the financial gains from automation are concentrated among executives, investors, and platforms. The risk is spread across the many; the reward accrues to the few.

THE MIRAGE OF RESKILLING

Tech leaders tout "reskilling" as a panacea. But the results are limited:

- Only **30% of displaced workers** in WEF pilot programs successfully transitioned to new roles [10].
- New jobs often require access to education, networks, and devices—resources not evenly distributed.
- Emerging roles like "prompt engineering" are niche and increasingly automated themselves.

Displacement is outpacing adaptation. And without a structural shift, no amount of training will rebalance the equation.

THE MYTH OF INEVITABLE PROGRESS

Automation is often framed as inevitable. But this is a narrative choice—not a law of physics.

These systems are designed by people, trained on human data, and deployed based on profit incentives. The automation avalanche didn't just happen. **We built it.**

And that means we can choose to build something else.

～

KEY TAKEAWAYS

- AI-driven automation is accelerating and now measurable across industries.
- Workers like Tanisha aren't just displaced—they're used as training data.
- Even when jobs remain, algorithmic management reduces agency and pay.

- Automation widens existing inequalities across race, class, and gender.
- Reskilling narratives often obscure the need for structural reform.

LOOKING AHEAD

In the next chapter, we'll follow the money. Who profits from this automation? How is user-generated data converted into billion-dollar revenue streams—and why do contributors see so little of it in return?

CHAPTER 4
THE PROFITS YOU NEVER SAW

THE $228 MYSTERY

EACH YEAR, Meta earns an average of **$228 per user** in Facebook-only advertising revenue in the U.S. and Canada, according to its 2023 Q4 earnings report [1]. *Multiply that by hundreds of millions of users, and the ad-driven tech economy becomes clearer: your attention is valuable. But not just your attention—your clicks, scrolls, likes, and comments are the raw materials of a massive prediction engine.*

And you're not just the target. You're also the source. You've trained the model.

They profit from your patterns. You profit from none.

∾

SURVEILLANCE CAPITALISM IN ACTION

The term "surveillance capitalism," coined by Shoshana Zuboff, refers to the monetization of personal data by tech companies. Every interaction is tracked, logged, and analyzed to predict and influence future behavior. This behavioral surplus becomes the fuel of monetization [2].

- **Google**: Over 80% of Alphabet's $320 billion in 2024 revenue came from advertising [3].
- **TikTok**: Its success lies in mining swipes, pauses, and replays to fine-tune recommendations.
- **Amazon**: Uses past purchase and browsing data not only to sell products—but to anticipate what you might want before you know it.

None of this value is returned to the people generating the data.

WHO REALLY PROFITS?

In the age of AI, it's not just engineers and executives who create value. It's users—through massive behavioral input. Yet the profits remain highly centralized:

- **OpenAI** projected over $2 billion in revenue for 2024, powered by training data scraped from public forums, user interactions, and web content [4].
- **Apple** positions on-device learning as a privacy win, yet still retains the value derived from local model fine-tuning—without redistribution [5].

These platforms sit atop mountains of behavioral data. And the contributors? They rarely even know they're contributing.

DATA WITHOUT DIVIDENDS

If you contribute to a commercial product, should you receive a share?

In most industries, labor and compensation go hand in hand. But the digital economy hides the labor—and hoards the rewards.

- You train voice assistants by correcting them.
- You refine AI vision systems by labeling images or solving CAPTCHAs.
- You teach recommender systems with every binge, skip, and like.

None of it is treated as labor. Why? Because acknowledging it would mean sharing the profits.

COMPARATIVE MODELS: ROYALTIES AND RENTS

Imagine if:

- Writers received royalties when their words were used in chatbot responses.
- Delivery drivers earned micro-dividends when their optimized routes became training data.
- Communities shared in the revenue when their language, culture, or routines trained predictive models.

These aren't pipe dreams—they reflect economic principles already in use:

- **Musicians** get paid per stream.
- **Landowners** receive royalties when resources are extracted.

- **Franchisors** get paid for usage of know-how and systems.

Why not digital laborers?

A GLOBAL INEQUITY: DIGITAL COLONIALISM

Much of the data training today's AI models comes from the Global South. Whether it's smartphone metadata, image tagging from crowd workers, or biometric verification in low-regulation zones—this data creates global value but rarely yields local benefit.

- Workers are underpaid or anonymized out of the value chain.
- Data is extracted under opaque terms from underrepresented communities.
- Local languages and cultures train global tools—but don't receive tailored benefits in return.

This asymmetry mirrors historic patterns of resource extraction. It's a modern form of **digital colonialism**.

TECH'S EVOLVING DEFENSE

Tech companies argue that data is aggregated, anonymized, or legally fair use. But legality does not equal fairness.

Copyright laws weren't designed for generative AI. Consent frameworks weren't made for invisible data trails. Platforms extract value at scale—while offloading the ethical and economic cost.

THE BROKEN FEEDBACK LOOP

A healthy economy rewards contributors. The data economy does not. The feedback loop—from value creation to reward—is broken.

- **Innovation** without reinvestment corrodes the commons.
- **Labor** without recognition creates resentment.
- **Profit** without distribution fosters instability.

This model isn't sustainable.
But it is fixable.

KEY TAKEAWAYS

- Tech companies monetize behavioral data at massive scale with no payout to users.
- Users train AI systems continuously and invisibly—without consent or compensation.
- Legal loopholes mask deep ethical failures.
- Models for sharing value already exist in other industries.
- Global power asymmetries reflect colonial patterns in digital form.

LOOKING AHEAD

In the next chapter, we'll define what a **Data Dividend** actually is. We'll explore the difference between selling your data and earning from your data—and how this shift could unlock a fairer, more participatory AI economy.

PART TWO
THE TECHNOLOGY OF ECONOMIC JUSTICE

CHAPTER 5
WHAT IS A DATA DIVIDEND?

FROM INVISIBLE INPUT TO STAKEHOLDER

IMAGINE THIS:

Every time you took a step with your fitness tracker, asked your smart speaker a question, clicked "like" on a photo, or chose a route on a delivery app—you earned a few cents. Not as a gimmick. Not through a shady app. But because you were recognized as a contributor to the AI economy.

Not a product.

Not a subject.

But a participant.

This is the radical premise behind the **Data Dividend***: the idea that if your data trains the machines, you should share in the value they generate.*

You power the system. Why not share in its success?

FROM EXPLOITATION TO PARTICIPATION

For decades, the digital economy has run on an asymmetrical deal: you use the platform, they take your data, they monetize it—often without your awareness or consent. This business model has powered trillion-dollar companies.

Behind the scenes, your digital footprints are more than residue. They include:

- **Behavioral signals** (clicks, scrolls, searches).
- **Sensor data** (location, biometrics, sleep patterns).
- **Creative contributions** (photos, comments, prompts).
- **Operational feedback** (delivery times, ratings, routing).

These become training data for AI systems. And yet, the labor embedded in this data is unpaid and unrecognized.

The Data Dividend flips that equation: if people power the product, people should share the profits.

DEFINING THE DATA DIVIDEND

A **Data Dividend** is:

A recurring compensation or benefit returned to individuals or communities based on the economic value generated from their personal, behavioral, or community-level data used in AI and automated systems.

This isn't about selling your privacy. It isn't about a one-time sale of your data, or a creepy marketplace for personal information. It's about building a new structure for recurring value distribution.

There are multiple models under the umbrella of a Data Dividend. Models include:

- **Individual payouts** (royalties for repeated contributions).

- **Collective dividends** (shared among platform users or communities).
- **Public trusts** (pooled value, distributed like Alaska's oil fund).
- **Equity models** (users as stakeholders in platform success).

Each model carries different tradeoffs for feasibility, scale, and equity. But they all share one premise: **data is labor—and labor deserves compensation.**

DATA IS THE NEW OIL—BUT IT'S YOURS

If data is "the new oil," as many have claimed, then we are not just passive bystanders. We are the oil fields.

When oil was discovered on private land, owners were granted royalties. Mineral rights and leasing arrangements became standard. If someone drilled on your land today and sold what they found, you'd expect payment.

So why not for data—when the "land" is your behavior?

The metaphor breaks down in one key way: **oil is finite**. Data regenerates every time you interact with a screen, device, or platform. That means the dividend is not a one-time event—it can be recurring, continuous, and scalable.

HOW A DATA DIVIDEND DIFFERS FROM DATA SALES

1. **Data Sales**
    - One-time transaction.
    - Often without transparency.
    - User loses control after transfer.
    - Based on volume.

2. **Data Dividend**
   - Ongoing participation-based payout.
   - Built on permission and traceability.
   - User retains governance rights.
   - Based on value and repeated utility.

A PROTOTYPE FROM CALIFORNIA

In 2019, California Governor Gavin Newsom proposed exploring a **"data dividend for Californians"**—an early attempt to surface the question of economic fairness in a digital age.

While the proposal stalled amid legal and technical complexity, it seeded new conversations:

- Could states negotiate on behalf of their citizens?
- Could unions or co-ops represent data contributors collectively?
- Should tech companies be obligated to share profits from user-derived AI?

Today, similar ideas are gaining traction globally—from blockchain-based data cooperatives (such as the **Streamr Network** in the EU, which lets users earn by sharing real-time data via decentralized protocols), to Kenya's **Ajira Digital** program, which promotes ethical digital labor and skills development. While not a data dividend per se, Ajira reflects a growing effort to center digital equity and fair compensation in data-rich economies.

BEHAVIORAL LABOR: INVISIBLE, UNPAID, ESSENTIAL

Let's break it down with examples:

- When you tag friends in photos, you're labeling training data for facial recognition.

- When you correct a voice assistant, you're refining a speech model.
- When you browse or linger, you train recommendation engines.
- When you wear a smartwatch, you contribute biometric data to fitness algorithms.

Each of these acts is a **micro-contribution**. Small in isolation. Massive in aggregate.

Today, these are extracted under the banner of convenience. But that convenience masks a deeper truth: these are forms of **behavioral labor**—valuable enough to power entire industries.

A COLLECTIVE OUTPUT, NOT A CORPORATE MIRACLE

Generative AI systems like ChatGPT, DALL·E, and Copilot aren't singular feats of engineering. They are **collective syntheses** of scraped text, code snippets, essays, artworks, reviews, stories, and more.

So why are we treating them as corporate marvels instead of human-powered collaborations? If we recognize these systems as collective outputs, then:

- Artists whose styles are mimicked,
- Programmers whose patterns are replicated,
- Writers whose tone is imitated,

...should be seen not as passive donors, but as **co-authors**. And co-authors receive royalties.

The Data Dividend simply extends that logic into the digital age.

WHY NOW?

Three converging forces make this the right moment:

1. **AI at Economic Scale**: These systems now generate billions in revenue.
2. **Ambient Data Generation**: Phones, watches, and homes constantly stream user behavior.
3. **Public Awareness**: People are increasingly aware that their digital lives are not just footprints—they are assets. However, this awareness is often shallow or misunderstood. Many still accept the notion that free services like search engines or social platforms are fair compensation for their data. This chapter aims to challenge that misconception—not just by highlighting the hidden value their data creates, but by raising the alarm: AI is no longer simply collecting data; it's using that data to displace the very people who generate it. Recognizing users as active contributors is not only a matter of justice—it's a matter of urgency.

As the AI economy expands, displaces labor, and concentrates wealth, the moral and economic case for data dividends grows stronger.

∼

KEY TAKEAWAYS

- The **Data Dividend** is a proposal to return value to the individuals and communities whose data fuels AI systems.
- This concept reframes digital interaction as **labor**, not just engagement.

- Dividends can take many forms: royalties, public trusts, collective payouts, or individual rewards.
- Without recognition and compensation, we risk deepening inequality in an already extractive economy.
- The time is ripe to begin building the legal, technical, and cultural infrastructure to support this vision.

LOOKING AHEAD

In the next chapter, we'll explore the infrastructure that makes this possible—from **confidential computing** and **smart contracts** to **personal data vaults**. These technologies shift power from platforms to people, and from surveillance to sovereignty.

CHAPTER 6
INFRASTRUCTURE FOR DATA DIGNITY

THE NAIROBI PILOT

DURING THE HEIGHT of the COVID-19 pandemic, a public health official in Nairobi joins a global data-sharing initiative. Hospitals around the world are uploading anonymized case data to help researchers spot patterns and predict outbreaks. But there's a dilemma: how can patient data remain private while still being useful for science?

The answer is a technology called **confidential computing**. In this illustrative scenario, Nairobi's hospitals process data locally inside encrypted environments called secure enclaves. The results—fully anonymized, fully computed—are shared without exposing underlying records. This strikes a balance: collaboration without compromise.

While fictional, this reflects real-world deployments, such as Fortanix enabling secure AI-driven analytics on sensitive patient data at UCSF and other medical centers [1] [2].

Security isn't just about hiding data. It's about enabling trust without exposure.

CASE STUDY: CONFIDENTIAL CLEAN ROOMS, CLEAR BOUNDARIES

In 2022, the Royal Bank of Canada (RBC) faced a challenge familiar to many institutions: how to personalize offers by combining transaction data with third-party merchant info—**without violating privacy**.

The solution? A **confidential clean room**.

Built on Microsoft Azure's Confidential Computing platform, this system let both parties run analytics in secure enclaves where neither could access the other's raw data. Insights were shared. The data wasn't. Trust wasn't assumed—it was cryptographically enforced.

This wasn't a pilot. It was production-grade infrastructure, showing that privacy and utility aren't mutually exclusive [3,4,5].

THE LIMITS OF PRIVACY-AS-SETTINGS

Today, digital privacy is often framed as a set of checkboxes:

- Turn off location tracking.
- Disable personalized ads.
- Read the terms and conditions (if you dare).

But these are illusions of control. Most people don't have time to parse legal documents. And even when they opt out, their data may still be used in aggregated or inferred ways.

True control doesn't mean reacting to extraction—it means authorizing it.

ENTER CONFIDENTIAL COMPUTING

Confidential computing refers to the ability to process encrypted data inside secure, tamper-proof hardware environments. These environments, called **trusted execution environments (TEEs)** or **secure enclaves**, protect data even while it's being used—even from the platform owner.

To simplify: imagine your data lives in a locked vault inside an apartment building. Only you have the key to open that vault—even the landlord (or the tech platform) can't access its contents. If someone wants to use your data, they can request a temporary key, but you decide when, how, and for how long they're allowed inside. The vault logs everything they do and slams shut when time is up.

Now, to show how this can feel seamless in real life: it's like unlocking your phone with Face ID. The phone knows it's you, but your face never leaves the device. A complex computation happens—but your private data stays private.

This means users could:

- Grant time-limited, purpose-specific access to their data.
- Participate in AI training without handing over raw files.
- Revoke access after the fact.

Confidential computing turns data from a one-way extraction into a two-way relationship.

FROM PASSIVE SUBJECTS TO DATA CITIZENS

Confidential computing enables new models of participation:

- **Data vaults:** Your information lives in a secure container that you control.
- **Zero-trust systems:** Even the platform itself can't peek into the data.
- **Permissioned AI:** Models can learn from you without remembering you.

This shift reimagines the user as an active participant—a **data citizen**—rather than a passive subject of surveillance.

SIDEBAR: BUILDING BLOCKS OF DATA DIGNITY

- *Component* — **Purpose.**
- *Confidential Computing* — Compute on data without exposing it.
- *Personal Data Vaults* — Store and permission data at the user level.
- *Smart Contracts* — Enforce access rules and rewards automatically.
- *Blockchain* — Provide a decentralized, tamper-proof ledger.

Together, these components form the **infrastructure of consent**—and will be explored in greater detail in the next chapter.

REAL-WORLD TECHNOLOGIES

Let's unpack a few technologies enabling these use cases:

- **Intel SGX:** Hardware-based trusted execution environment that allows code to run securely on sensitive data [6].
- **Arm TrustZone:** Built into billions of smartphones, it

segments trusted operations—used for Face ID, secure payments, and now increasingly for machine learning [7] [8].
- **Azure Confidential Computing**: Microsoft's cloud platform offers virtual machines with built-in secure enclaves, enabling enterprises to move sensitive workloads into the cloud without exposure [9].
- **Apple Secure Enclave**: Powers Face ID, Touch ID, and local AI processing [10].

And in the public sector:

- **RBC's Virtual Clean Room** allows banks to compute on shared datasets without disclosing customer information [11].
- **Estonia's e-Governance** uses decentralized identity and encrypted channels to give citizens ownership of their records [12].

These tools don't just protect privacy—they create the conditions for trustworthy collaboration.

WHAT THESE USE CASES TEACH US

1. **It scales**: Confidential computing is not theoretical. It runs on consumer smartphones and enterprise clouds.
2. **It's cross-sector**: From finance to health to advertising, any domain that touches sensitive data stands to benefit.
3. **It's invisible to the user**: Most people don't know their iPhone's Neural Engine uses a secure enclave. And that's the point—it works silently, without burdening the user.

HISTORIC PARALLELS: THE SECRET BALLOT

In the 19th century, the secret ballot revolutionized democracy. It allowed people to express their will without fear of coercion.

Confidential computing is the digital equivalent. It lets people contribute to public or commercial systems—AI, research, governance—without giving up control or exposing themselves.

You participate. But you remain protected.

WHAT'S MISSING

Despite its growing use, confidential computing hasn't yet been applied at scale to the data labor economy. Why?

- **Lack of user-facing tools**: Most deployments are enterprise-facing. Individuals don't yet have vaults or dashboards.
- **No standard for value exchange**: While computation is secured, compensation isn't built into the architecture.
- **Limited awareness**: People don't know their data could be protected—and paid for—at the same time.

The challenge isn't technological. It's institutional, economic, and cultural.

THE BRIDGE TO DATA DIGNITY

Confidential computing isn't a complete system—it's a set of powerful building blocks. While already trusted in sensitive domains like finance and healthcare, its role in enabling a fair data economy is still emerging. Like early electrical grids or the foundational protocols of the internet, this technology offers the scaffolding—but society must decide what to build on top.

We could build:

- **Personal data vaults** that use secure enclaves.
- **Marketplaces** where users decide when and how to share data for AI training.
- **Smart contracts** that release payments when data is used under agreed conditions.

These systems don't require fantasy breakthroughs. They require coordination and the political will to shift power.

KEY TAKEAWAYS

- Confidential computing allows data to be processed securely—even while in use—without exposing it, not even to the platform or service provider.
- It's already used in production environments across finance, healthcare, and mobile ecosystems.
- It empowers users to participate in AI systems without giving up control of their data.
- The technology is mature, but user-facing tools and economic models are lagging behind.
- Confidential computing offers the scaffolding for a more equitable data economy—if combined with user control, attribution, and compensation.

LOOKING AHEAD

Confidential computing is just one piece of the puzzle. In the next chapter, we'll look at what happens when privacy-preserving infrastructure meets user power—through tools like **personal data**

vaults, **smart contracts**, and **blockchain**. These technologies *can* move us beyond protection and toward true participation, helping users define the terms of engagement in a data economy built on consent and control.

CHAPTER 7
YOUR DATA, YOUR RULES

A CHOICE THAT WASN'T

WHEN JAMAL SIGNED up for a fitness tracking app, he scrolled through the 19-page terms of service like everyone else—clicking "agree" without reading a word. He just wanted to monitor his heart rate and track his morning jogs. A year later, he saw an ad for a high-risk insurance plan that eerily referenced his fitness habits. The data broker who sold his activity patterns to an underwriter? Legally in the clear. Jamal? Out of options.

> This wasn't consent. It was compliance masquerading as choice.

∽

THE PROBLEM WITH PERMISSION

In theory, we live in a permission-based digital world. But in practice, "consent" often means clicking through confusing interfaces, vague language, and take-it-or-leave-it terms. Even when platforms offer privacy settings, they're often buried or difficult to understand.

And even when you revoke permission, the data trail remains—often sold, shared, or analyzed without further notice.

This is not informed consent. It's a loophole system that benefits data extractors, not users.

WHAT REAL CONTROL LOOKS LIKE

Real data control means:

- **Granular consent**: You choose what type of data is shared, for what purpose, and for how long.
- **Revocability**: You can change your mind and take your data back.
- **Transparency**: You see who has access to your data and what they do with it.
- **Auditability**: A clear log tracks every use of your data.

These are the core principles of emerging **permissioned data systems**—architectures designed not just to protect data, but to give you agency over it.

PERSONAL DATA VAULTS

Think of a personal data vault like a safety deposit box for your digital self. Instead of spreading your information across countless servers, apps, and clouds, you store it in a single, encrypted location. Apps must request access, and you decide whether to grant it.

Examples in the wild:

- **Solid by Inrupt**: A project led by web inventor Tim Berners-Lee, giving users a "pod" where their data lives, controlled by them [1].
- **MyData Operators**: A network of organizations in Europe developing human-centric data intermediaries [2].

Instead of being tracked and targeted invisibly, you participate visibly and voluntarily.

SMART CONTRACTS FOR SMART DATA

Imagine if you could grant someone temporary access to your digital door lock—like handing out a time-bound, self-expiring key—and that key automatically vanished after a set amount of time. That's the idea behind **smart contracts**: self-executing digital agreements that do what they say, without needing human intervention or middlemen.

Built on **blockchain**—a secure, decentralized system of record-keeping—smart contracts allow users to enforce data permissions automatically. Here's how it works in practice (one example among many possible domains):

- You want to contribute your medical data to a research study—but only for six months.
- A smart contract is programmed with your condition: access begins today and ends in 180 days.
- Once the time is up, the contract shuts access off—no one needs to remind it.
- Every time your data is accessed, it's logged on the blockchain in a permanent, transparent way.

Think of it like giving someone a guest pass to a coworking space. The pass automatically deactivates after one day and logs exactly who used it, what time they entered and exited, and which conference rooms or equipment they accessed. It's not just access—it's accountable access.

These systems aren't just theoretical. Projects like **Ocean Protocol** and **Mina Protocol** are pioneering smart contract infrastructure for ethical data sharing in areas like climate science and health research.

The promise: your data, your terms—enforced by code you can trust.

THE SHIFT FROM PASSIVE TO PERMISSIONED

Most of today's platforms are built for extraction: get the data, lock the user in, sell the insights.

But a permissioned model flips the script. It assumes:

- The user owns their data.
- Platforms must earn trust to access it.
- Value must be shared with contributors.

This isn't a minor tweak to user experience (UX). It's a different business logic—one that requires legal frameworks, technical infrastructure, and cultural expectations to evolve together.

CHALLENGES ON THE PATH

Building a permissioned data economy isn't easy:

- **Interoperability**: Platforms must adopt shared standards for data access and revocation. For example, if your health data vault uses a different format from your fitness app, it's like two people speaking different languages. Interoperability ensures that data can flow smoothly between systems, reducing friction for users and developers.

- **Incentives**: Many companies profit from hoarding data. Why would they change? For companies to adopt permissioned models, they need to see benefits, such as higher trust, better quality data, or competitive advantage. For example, Apple markets privacy as a feature to attract users.
- **Simplicity**: Complex permissions must feel effortless for users. No one wants to micromanage every click. For example, when Apple introduced Face ID, it made advanced security feel seamless to everyday users.

Still, we've seen similar transitions before:

- **From cash to digital wallets:** interoperability between banks, card networks, and payment apps allowed mobile payments to take off.
- **From landlines to mobile networks:** telecom companies had to create incentives for users to adopt, like cheaper plans, and ensure simplicity with universal SIM cards and standards.
- **From closed file formats to open standards:** Microsoft Office adopting open XML formats allowed documents to be read and edited across platforms, enabling interoperability, incentivizing collaboration, and simplifying user workflows.

The tools are here. What's needed is the will.

KEY TAKEAWAYS

- Today's digital consent is often superficial—true control is rare.

- Permissioned data systems give users real agency: who sees what, when, and why.
- Personal data vaults and smart contracts can enforce consent in practice.
- The path forward demands infrastructure, policy, and cultural change.

LOOKING AHEAD

Next, we'll explore a provocative idea: that data isn't just something you *own*, but something you *produce*—and that production is labor. What would it look like to meter, measure, and compensate the invisible work of your digital life?

CHAPTER 8
DATA AS LABOR, TRACKED AND PAID

THE INVISIBLE SHIFT

PRIYA WORKS *as a freelance graphic designer. She gets paid when she submits a design, tracks her hours, and invoices her clients. One day, she tries an AI design tool and is amazed—it generates templates in seconds based on user preferences. Then she learns the shocking part: the tool was trained on millions of images, including design portfolios like hers, scraped from the web without permission or pay.*

She realizes she's been working all along—just not for herself. Her creativity fed a system that never asked, never credited, and never compensated.

We produce data constantly, but we only get paid for our labor when it's visible.

∽

THE UNSEEN ECONOMY

Every time we search, click, scroll, rate, or upload, we generate data. These digital footprints—what economists might call behavioral exhaust—are harvested by companies, structured into datasets, and used to train or optimize powerful AI models. This ongoing stream of human activity underpins predictive algorithms, personalized services, and automation strategies across industries.

Consider a few examples:

- A teenager rates songs on a music app, helping fine-tune its recommendation algorithm and engagement metrics. This data not only personalizes her experience but also contributes to a dataset that improves the platform's predictive models across millions of users, thereby driving ad revenue and subscription growth.
- A delivery driver's GPS trails help their employer optimize delivery logistics and driver efficiency, while also feeding aggregated datasets used by mapping platforms and AI developers to refine traffic prediction systems.
- A support worker answers customer chats, which are logged and analyzed to train natural language models that will eventually automate large portions of support operations, improving efficiency but reducing demand for human agents.

To the user, these are routine activities. To the platform, they are unpaid contributions that improve core products and reduce long-term costs. The data derived from these actions generates measurable economic value—yet none of it returns to the individuals who produced it.

This dynamic creates an asymmetry: a handful of companies convert collective behavior into private profit, while the very people who fuel the system remain invisible in its economics. It's a data economy, yet only a select few get paid.

The result? A two-tiered system: companies and platforms harvest the value of AI, while individuals—the users whose behavior, content, and interactions train these systems—continuously subsidize it, often without knowing.

HOW BIG IS THIS ECONOMY?

It's difficult to pinpoint the value of a single scroll, tag, or photo upload—but the aggregate economic impact of human-generated data is massive. To appreciate the scale of the unseen economy, we can examine how digital platforms monetize behavioral data:

- Meta earns an average of $228 per user annually in the U.S. and Canada, primarily through ads personalized using behavioral data [1]. Multiply that by hundreds of millions of users, and the result is tens of billions in annual revenue driven by user activity.
- Stack Overflow posts and web forum content helped train GitHub Copilot and ChatGPT—AI tools now generating substantial subscription income. Copilot costs $10–$30/month per developer. ChatGPT offers paid tiers starting at $20/month for individuals and scales up to enterprise-level pricing. These tools wouldn't function without massive volumes of unpaid user-generated content.
- Instacart, DoorDash, and Uber used gig workers' location and behavior data to optimize delivery routes and forecast demand. These patterns were then embedded in automation strategies, leading to investments in robotic delivery pilots [2] and self-driving freight [3]. Gig workers helped train the systems that may one day replace them,

without seeing a share of the long-term value they created.

This is the unseen economy in action: vast, profitable, and powered by human contributions that remain uncompensated. Data is labor, but the labor force is invisible in the revenue model.

WHAT IS DATA LABOR?

When you click, tag, label, search, type, or even pause your scroll, you're producing signals that machines interpret and use to learn. This behavioral exhaust is not just metadata. It is labor—not necessarily in the sense of strenuous effort, but in the economic sense of producing value that would typically justify compensation. Think of mowing someone's lawn, painting their fence, or delivering groceries: these are not grand feats of effort, but they are paid because they produce recognizable, useful outcomes. Likewise, digital actions like labeling images or correcting grammar produce outputs that train AI and improve automated systems. They generate value that resembles paid labor, and they should be recognized as such. This is the essence of what we call **Data Labor**—the recognition that digital actions which generate value deserve acknowledgment and potentially compensation, just like any other form of labor.

To understand what makes data labor so different, consider how traditional labor is structured: it's visible, formal, and compensated. You clock in, perform a task, and receive wages in return. Data labor, by contrast, happens quietly and continuously through digital interactions. It's hidden in plain sight—unacknowledged, uncompensated, and often misunderstood as mere online activity. Recognizing it as labor is the first step toward ensuring that the people who generate value online also share in the rewards.

DATA AS LABOR: A FRAMEWORK

To rebalance the equation, we need a new lens: **data as labor**. This concept frames our digital activity not as ambient noise, but as valuable work that powers AI.

Three core principles make this model function:

- **Attribution**: This means being able to trace specific pieces of data—such as a product review, a map correction, or an annotated photo—back to the person who contributed it. Without attribution, value disappears into the platform with no way to credit the source. Attribution is the foundation of recognizing who did the work.
- **Consent**: True participation requires permission. Consent means individuals should have the ability to decide when, how, and for what purpose their data is used. Right now, consent is often buried in terms of service. In a data labor model, it becomes active, ongoing, and revocable.
- **Compensation**: If your contribution creates value, you should receive a share of it. This doesn't necessarily mean every click is worth a dollar, but that there should be systems for fair distribution—similar to how musicians earn royalties when their songs are streamed. If your data helps train a model that's later monetized, a portion of that value should come back to you.

Consider an example from the world of photography. If a stock photo you uploaded to a licensing platform is used in a marketing campaign, you are credited and paid accordingly. You gave permission, your name is attached, and you're compensated when your work generates value. That's attribution, consent, and compensation working together. The same logic should apply to data labor. If your online contributions help train a system or refine a product, you should be able to see where your input went, say yes or no to its use, and share in the resulting benefits.

EMERGING MODELS AND EXPERIMENTS

While the concept is still young, early-stage efforts are underway—offering glimpses of what a future data labor economy might look like in practice.

- **Spice AI** and **RepubliK** are developing systems that reward users with tokens for their data contributions. In these models, users can choose to share specific types of data (like their activity on certain platforms or preferences in media consumption) and receive digital tokens in return. These tokens can sometimes be exchanged for services, access, or even money, depending on the platform's economy [4][5].
- **Data unions** like **Swash** and **Streamr** aim to pool individual user data into collective bargaining units. Members of these unions opt in to share their data through browser extensions or apps, which then aggregate it securely. The unions negotiate with data buyers—typically marketers, researchers, or developers—and distribute revenues back to the members proportionally. Swash, for instance, allows users to earn SWASH tokens for sharing browsing data and has integrated with projects like Ocean Protocol to broaden its reach [6][7].
- **OpenMined** provides a different but complementary approach, focusing on privacy-preserving tools that allow data to be used for machine learning without ever exposing the raw data itself. Their open-source technologies support federated learning and differential

privacy—enabling data owners to contribute to AI development while maintaining control over what is shared. This supports the principle of consent by giving contributors granular control over their data [8].

These projects are imperfect and still evolving, but they each bring the core principles of attribution, consent, and compensation into practice. By experimenting with ways to trace, reward, and protect data contributions, they offer early blueprints for a more equitable digital economy—one where users are seen not just as passive participants, but as contributors deserving of recognition and reward.

WHAT'S MISSING ON THE PATH TO A DATA DIVIDEND?

The challenge isn't just technology. It's economic, legal, and cultural:

- **No pricing standard for data labor:** In traditional labor markets, tasks are priced based on time, skill, or output. But when it comes to digital activity—like correcting AI-generated text or uploading labeled photos—there's no common framework to value these contributions. Without pricing benchmarks, there's no way to measure how much value was created or to justify payment. A pricing standard for data labor would require collaborative efforts to define units of contribution and tie them to the value they help create, much like the way royalties are set for creative work.
- **Limited tools for tracking attribution:** Even when valuable data is contributed, most systems don't track where it came from or how it was used. Data gets pooled, stripped of context, and loses its connection to the person who generated it. This makes it nearly impossible to credit or compensate individuals. New tools—like audit

logs, cryptographic tags, and federated learning models —could help preserve that link from contributor to outcome.

- **Weak incentives for platforms to share value:** The current business models of major tech companies are built on maximizing data extraction without having to pay for it. Sharing value would mean sharing power, and that goes against their commercial logic. Changing this will require not just new tools, but new policies—ones that mandate transparency, require revenue sharing, or reward companies for ethical data stewardship.

Just as labor movements reshaped industrial capitalism—securing rights like the minimum wage, collective bargaining, and workplace safety—a new movement may be needed to make data work visible, valuable, and fair. Early signs are already emerging:

- The **AI Now Institute** is a nonprofit research organization dedicated to studying the social implications of artificial intelligence, with a focus on developing policy frameworks for algorithmic accountability, transparency, and the protection of labor rights in the age of automation [9].
- The **DataUnion Foundation** develops tools and standards to enable collective bargaining over personal data through Data Unions, creating transparent systems for individuals to pool and monetize their data while participating in fair revenue-sharing models [10].
- **MyData Global** promotes human-centric data governance and has developed the MyData principles for ethical data handling, transparency, and user control [11].
- **Fairwork Foundation** evaluates gig economy platforms against principles of fair pay, conditions, contracts, management, and representation—offering a framework

to assess and improve labor standards in digital work environments [12].

- **The Self-Employed Women's Association (SEWA)** in India has pioneered models of collective power for informal workers—many of whom were historically excluded from labor protections. Their cooperative approach offers lessons in how distributed labor forces can self-organize for recognition and rights [13].

But history reminds us: movements only succeed when those doing the work organize and demand change. Data labor is invisible today—but it doesn't have to be tomorrow.

KEY TAKEAWAYS

- Most AI systems are built on uncompensated human data.
- The idea of treating "data as labor" reframes this as a matter of economic fairness and rights: those who generate valuable digital activity should be recognized as contributors, not merely consumers.
- Attribution, consent, and compensation provide the scaffolding for a fairer model—ensuring that contributions can be traced, permission is secured, and value is shared when data is used to generate profit.
- A data dividend—sharing the wealth generated by AI with those who help build it—is feasible, but it will require collective action and structural support across technology platforms, legal frameworks, and social norms.

LOOKING AHEAD

If data is labor, then the next question is delivery: how do we move from this recognition to actual dividends? What infrastructure, institutions, and incentives are required to turn this economic value into income people can see and use? In the next chapter, we ask whether universal basic income (UBI) is enough—and explore how a data dividend model can offer not just support, but agency.

PART THREE
BEYOND UNIVERSAL BASIC INCOME

CHAPTER 9
UBI ALONE ISN'T ENOUGH — WHY DIGNITY BEATS DEPENDENCY

A DIVIDEND BY DESIGN

IN A SMALL TOWN outside of Helsinki, a group of residents participate in two programs. One is a well-regarded UBI pilot funded by the Finnish government [1]. The other is a municipal initiative that collaborates with Finland's **DataMust** pilot, a real-world project led by VTT and supported by Nokia [2]. This initiative explores edge-based data marketplaces, where local residents and small businesses can voluntarily share data—such as mobility patterns or energy usage—for community and commercial value.

The UBI helps stabilize their lives. It reduces financial stress and opens up time for caregiving, retraining, or creative pursuits. But the data-sharing platform offers something different: a sense of authorship. Participants choose what data to share, understand its use cases—like improving traffic flow or enabling smarter infrastructure—and know that they are contributing to civic innovation.

Nobody here sees UBI as a handout—it's part of a thoughtful social safety net. But the data dividend feels participatory. It reframes people not just as recipients of public generosity, but as contributors to a new digital economy. And that difference, for many, feels profound.

. . .

The UBI brought relief. The data dividend brought recognition.

THE PROMISE AND LIMITS OF UNIVERSAL BASIC INCOME

Universal Basic Income (UBI) is a policy concept that proposes regular, unconditional payments to all individuals regardless of employment status or income level. The idea is to provide a financial floor that ensures basic economic security, allowing people to meet essential needs without bureaucratic hurdles or work requirements.

UBI has its roots in centuries of economic thought but gained renewed attention in the 20th and 21st centuries, particularly in response to rising inequality and the growing impact of automation. It has been tested in various forms around the world, including:

- The **Alaska Permanent Fund**, established in 1976, invests oil revenue on behalf of state residents. Every year, eligible Alaskans receive an annual dividend—ranging from several hundred to over a thousand dollars—regardless of income or employment status. This model demonstrates how resource wealth can be shared directly with the public [3].
- **Finland's national UBI pilot** (2017–2018) provided 2,000 unemployed individuals with a monthly payment of €560, with no strings attached. Results showed improved mental well-being and higher life satisfaction among recipients, even though employment rates saw modest changes [4].

- Local trials in **Canada** (Ontario's Basic Income Pilot, 2017–2018)[5] and the **United States** (Stockton Economic Empowerment Demonstration, 2019–2021)[6] offered unconditional monthly stipends. In Ontario, recipients reported better physical and mental health, and more stable housing before the program was canceled. In Stockton, participants experienced reduced income volatility and increased full-time employment.

These experiments have shown that UBI can:

- **Reduce poverty and financial stress**: In the Stockton pilot, participants reported greater income stability and a decreased reliance on payday loans. Those receiving the basic income were more likely to pay off debts and cover unexpected expenses, offering real relief from chronic financial pressure.
- **Improve health and mental well-being**: Finland's national UBI trial documented improved mental health and higher life satisfaction among participants, even though employment outcomes remained largely unchanged. Similarly, Ontario's pilot found that recipients experienced less anxiety and better overall health during the program's run.
- **Enable individuals to pursue education, caregiving, or creative work without fear of destitution**: In Ontario, some participants went back to school or cared for elderly family members, while others used the breathing room to start small businesses. This shift in how people chose to spend their time highlights UBI's potential to unlock human potential beyond basic survival.

At its best, UBI offers breathing room. It gives people time to transition during periods of economic upheaval—such as those brought on by automation and artificial intelligence—while preserving their dignity and autonomy.

But even UBI advocates acknowledge its limits. It offers support, but not participation. It ensures survival, but not agency. And structurally, it remains a top-down solution: companies are taxed, government officials decide how much each individual receives, and those decisions are vulnerable to political winds. The result can be a system where recipients are tethered to the ideology of the governing party—relying on continued support rather than shaping their own economic role. Without structural changes to who owns and benefits from AI systems, UBI risks becoming a band-aid for deeper power imbalances. A data dividend model, by contrast, is inherently bottom-up: users generate the value, consent to its use, and are compensated directly for their contributions.

WHY A DATA DIVIDEND COMPLEMENTS UBI

A data dividend doesn't replace UBI. It complements it.

Where UBI is universal and unconditional, data dividends are **earned and participatory**. They acknowledge that users are already contributing to the digital economy—not through traditional labor, but through data that trains and fuels AI systems.

Importantly, the promise of data dividends may influence user behavior and decision-making in ways that extend beyond passive participation. If people know their data contributions will be valued and compensated, they may begin to make choices accordingly: selecting smart home devices, vehicles, or apps not just for convenience, but for their potential return on data value. Someone might choose an electric vehicle known to yield high-value data, or install smart security cameras with an eye toward contributing to community safety datasets. This opens a new frontier of user agency—where economic participation and digital sovereignty become intertwined.

By pairing these two models, we get the best of both:

- **Stability from UBI**, which provides a predictable financial baseline. This gives individuals the confidence to weather job transitions, invest in retraining, or pursue caregiving and creative work—without the fear of losing everything. It acts as a modern safety net, making economic disruption less devastating.
- **Recognition through data dividends**, which shifts individuals from being passive subjects of data collection to acknowledged participants in the AI economy. This recognition reinforces dignity by valuing people's contributions and granting them a stake in the digital wealth they help generate.

Together, they shift the narrative from dependency to participation—blending foundational support with active economic inclusion.

EXAMPLES OF PARTICIPATORY COMPENSATION

The idea of compensating people for their contributions to broader systems of value creation isn't unprecedented:

- **Music and film royalties** compensate creators when their work is reused. As discussed earlier, musicians receive payments when their songs are streamed, sampled, or broadcast. This establishes a clear precedent for recurring value generated from original content—even if it's consumed repeatedly by others.
- **Loyalty programs** reward consumer behavior with points, discounts, or perks. Consider frequent flyer programs: the more you fly, the more points you earn, which can be redeemed for upgrades, flights, or elite status. These programs treat your engagement as

valuable, and they reward you for it—much like data dividends would do with everyday digital interactions.
- **Crowdsourced innovation platforms** such as GitHub and Kaggle compensate users based on their contributions. Developers who maintain open-source codebases gain reputation and sometimes direct sponsorship from companies that rely on their work. Data scientists competing in Kaggle challenges can earn prize money or career opportunities based on the performance of their models.

In each of these cases, value is recognized and shared with the contributor. Why not apply the same logic to data—the digital resource shaping the 21st-century economy?

ANTICIPATING CRITIQUES

Some worry that data dividends won't be large enough to matter. Their skepticism stems from several sources. To take these concerns seriously and respond with clarity, we'll address each of the major critiques one at a time—examining the assumptions behind them and offering counterpoints rooted in economics, technology, and social precedent. First, critics often point out that most users produce data in ways that are difficult to value individually—clicks, scrolls, location pings, and usage patterns may seem insignificant on their own. Without a clear market price for such micro-interactions, it's hard for skeptics to imagine how these behaviors could yield meaningful returns.

Second, opponents question whether the data economy can be redistributed at all. They argue that the infrastructure for identifying, attributing, and compensating individual data contributions does not yet exist at scale. From their perspective, the complexity of tracking which data trained what AI model—and how much that training is worth—makes compensation nearly impossible to operationalize.

Others view the issue as economic naiveté. They worry that the cost of paying millions or billions of users would erode corporate profits and innovation incentives. Some critics also fear a regulatory overreach that could stifle growth in fast-moving industries, especially if companies are required to audit and disclose detailed data flows. These concerns, while often sincere, tend to underestimate the pace of innovation in tracking, encryption, and smart contract technologies.

Finally, there's an ideological barrier. Some critics see data sharing as a voluntary trade for free services, not a form of labor. From this perspective, the user has already been "paid"—in access to platforms like social media or cloud storage—and asking for more seems unjustified. Yet this view ignores how value continues to accumulate and compound on top of user contributions long after the initial interaction.

To engage these critiques thoughtfully, we'll address each one individually in the sections that follow—beginning with concerns about the value of user data, and moving through technical, economic, and ideological objections. Each deserves a fair hearing—and a strong response.

RESPONSE TO CRITIQUE #1: "USER DATA IS TOO SMALL TO MATTER"

The concern that individual user data is too insignificant to justify compensation may seem intuitive—but it doesn't hold up to scrutiny when you look at the full picture. Tech companies like Meta have reported making over $228 in annual advertising revenue per user in the U.S. and Canada alone, according to Meta's Q4 2023 earnings report [7]. And Meta is just one company.

Amazon, for example, generates billions through its ad business, fueled by the behavioral signals of shoppers [8]. Google's ecosystem—

spanning search, YouTube, Gmail, and Android—yields even more, with Alphabet reporting over $300 billion in ad revenue in 2023 [9]. TikTok, X (formerly Twitter), and LinkedIn all rely on user-generated data to target ads and improve content delivery. Even Apple, while more privacy-focused, has recently expanded its advertising business by leveraging first-party data [10].

When you begin to add these up, the numbers become staggering. If even a fraction of this value were routed back to users—through transparent systems that tracked, priced, and attributed data usage—data dividends could rival traditional wages. A household of four active internet users could, over time, receive hundreds or even thousands of dollars per year. That's not just pocket change. That's rent. That's groceries. That's a foundation.

And advertising is only part of the story. As companies deploy AI agents to replace or augment human labor, they unlock a second engine of profit. These systems—customer service bots, code generators, copywriters, and beyond—lower operational expenses by doing what once required a salary, a schedule, and a benefits package. Unlike human workers, these agents don't sleep, don't unionize, and don't need onboarding. They are tireless, scalable, and increasingly productive.

This means the margin between what a company earns and what it pays out grows even larger. And that growing margin is made possible in part by the data that trained these agents—data produced by real people. The more companies rely on these systems to replace human labor, the more compelling the case becomes for recognizing and rewarding the original source of that labor.

When data is used not just to sell ads but to fuel automation itself, the pool of value available to share expands dramatically. The critiques that data dividends are too small begin to feel shortsighted. If structured correctly, this could be a paradigm shift—one that offers not just dignity, but real income, even salary-replacement wages, for contributing to the digital economy.

RESPONSE TO CRITIQUE #2: "THE INFRASTRUCTURE DOESN'T EXIST"

Another common critique of data dividends is the belief that the infrastructure for fairly compensating individual contributors does not exist—or is too complex to scale. But we've seen precedent for distributed value-sharing models, even with limited tools.

Data dividends are already part of our economic fabric—just not always labeled as such. Consider frequent flyer miles. When airlines began facing pressure to retain customers and fill seats, they didn't slash prices. Instead, they created loyalty programs: points for miles flown, redeemable for future travel, upgrades, or elite perks. These rewards weren't paid in cash, but they carried real value. And they were powered by the data airlines collected about travelers—their routes, frequency, and preferences.

Frequent flyer programs emerged in the 1980s, long before the digital data economy as we know it took shape. They were designed with rudimentary tools and analog systems of accounting and tracking. Yet they proved that sustained, trackable customer engagement could be converted into a form of economic return.

Today, we have exponentially more sophisticated tools—blockchains, confidential computing, smart contracts, personal data vaults. These tools enable the secure attribution, pricing, and exchange of data while protecting user privacy and enabling consent. The barriers are no longer technological; they are institutional and conceptual.

We are in the early phases of adapting these technologies to serve the architecture of the data economy. The dividends that emerge may begin as points, tokens, or micropayments—but they are stepping stones to something larger. What's needed now is the will to build interoperable, user-facing systems that make these possibilities tangible.

Waiting for perfection only delays progress. Meanwhile, the rate of job displacement accelerates, and wealth inequality deepens. If we don't begin building the pipes for a data labor economy now, we risk hardening the existing imbalance and leaving people permanently excluded from the value they help create.

RESPONSE TO CRITIQUE #3: "IT WILL HURT CORPORATE PROFITS"

Another objection raised by skeptics is that compensating individuals for their data would erode corporate profits and stifle innovation. From this perspective, tech companies have little incentive to support such a system—why would they share a slice of the revenue pie they've long claimed entirely for themselves?

This critique assumes a zero-sum view of economic participation. But what if shifting toward more equitable data sharing is not just ethically sound, but strategically smart? What if the long-term viability of tech platforms depends not on maximizing short-term margins, but on building deeper, more trusting relationships with users?

These are the questions we'll explore in the next chapter. We'll examine why a data dividend model may actually benefit tech companies—helping them attract better data, retain user loyalty, and avoid regulatory fallout. We'll look at the early adopters who are already experimenting with participatory models, and explain why forward-thinking companies are beginning to view trust as an asset —not a liability.

In short, we don't have to rely on idealism to imagine corporate cooperation. We can point to business logic. And we'll make that case in the next chapter.

RESPONSE TO CRITIQUE #4: "YOU'RE ALREADY PAID WITH FREE SERVICES"

The final critique is rooted in ideology: the belief that data sharing is voluntary and users are already compensated through free services like social media, email, or cloud storage. From this perspective, asking for additional compensation is seen as excessive—after all, the deal was struck when you clicked "I agree."

But this view misunderstands both the nature and the scale of what's being given away. Free services may offer short-term utility, but they extract long-term value that continues to compound. The algorithms improve, the data pools grow, the predictions sharpen—and the platforms profit handsomely, often without further consent or awareness. Meanwhile, users continue generating new data with every interaction or click, yet remain largely excluded from meaningful control or compensation.

Data dividends challenge this outdated bargain. They ask us to rethink the status quo—not just as consumers accepting free tools, but as contributors deserving recognition and a fair share. This isn't about taking down the platforms. It's about recalibrating the relationship.

And at its heart, this isn't just a moral argument—it's a call to shift the norms that define digital participation:

- **Data use becomes consensual**: Instead of burying permissions in legalese, users can explicitly authorize how their data is used, for what purposes, and for how long.
- **Value generation becomes transparent**: Individuals can see when and how their contributions create economic value, just as one might track royalties or dividends in other industries.
- **Users become partners in value creation**: Rather than being tracked invisibly or treated as raw material, users are acknowledged as co-creators in the systems their data helps to build.

Crucially, this model need not compromise privacy. Emerging tools like confidential computing, personal data vaults, zero-knowledge proofs, blockchain, and smart contracts make it possible to build systems that are both secure and respectful—where individuals don't have to choose between dignity and convenience.

The norms we set today will define the digital economy of tomorrow. Building models that respect users and reward contributions isn't just fair—it's foundational to a sustainable, trustworthy future.

A SOCIAL CONTRACT THAT REFLECTS OUR TIMES

Before the factory age, we lived in a world defined by feudal hierarchies and agrarian sharecropping—systems where power and wealth were concentrated in the hands of a few, and most people toiled without rights or recourse. There was little to no compensation for labor, no protections for health or well-being, and certainly no say in how the value they created was distributed.

The industrial revolution disrupted that world, and over time, we built new institutions to match the new economy:

- **Wages for labor**: Workers once bartered or received room and board instead of pay. The recognition that labor should earn money—real, negotiable wages—was a moral and economic breakthrough. It transformed workers into independent economic agents and made labor a centerpiece of modern capitalism. Securing this right took decades of organizing and resistance.
- **Laws against child labor and unsafe working conditions**: Early industrial factories often employed children and exposed all workers to brutal, dangerous conditions. It took tragedy, outrage, and the rise of labor movements to push for legal protections. These laws asserted that human life and dignity were not disposable inputs.

- **The 40-hour work week and the weekend**: Born from grueling labor battles, the idea that workers should not be expected to work endlessly led to the standard work week. It wasn't gifted—it was won through strikes and protests. The weekend enabled rest, family life, and civic participation.
- **Unions for collective bargaining**: Individual workers had little leverage. But together, they negotiated better wages, safer conditions, and benefits. Unions institutionalized the idea that workers deserve a voice in the systems that govern their work.
- **Social safety nets like pensions and workers' compensation**: The understanding that injury, aging, or bad luck shouldn't plunge people into destitution led to policies like unemployment insurance and retirement benefits. These safeguards recognized workers as long-term contributors to national prosperity—not disposable parts.

These changes weren't handed down from above. They emerged—through organizing, struggle, protest, and policy reform. It took decades to align our social contract with the realities of industrial labor.

We are at a similar juncture now. The factory economy is becoming less relevant as AI and automation increasingly displace traditional jobs. Yet the institutions that govern value creation and distribution haven't caught up.

In the digital age, we must build new institutions around **data**—the raw material of AI. That means:

- **Recognizing that data is labor**: Just as the physical labor of the industrial era produced goods, our digital behaviors generate value. This recognition reframes participation in digital life as economically meaningful.

- **Creating systems for consent, compensation, and accountability**: These are the pillars of a fair data economy. Consent ensures agency, compensation ensures fairness, and accountability ensures that data is not misused or exploited without consequence.
- **Ensuring that wealth created by AI reflects the contributions of the many, not just the few**: The benefits of AI must be distributed—not hoarded. That means building pipelines that route value back to contributors, just as royalties reward creators or dividends reward shareholders.

This isn't about utopia. It's about fairness. It's about dignity.

And it's about understanding that, just as the industrial revolution required new laws, rights, and institutions to humanize the factory floor, the digital revolution demands new protections and new opportunities to humanize the data stream.

THE PATH FORWARD

A data dividend model paired with UBI is not just a policy proposal—it's a framework for a more participatory, equitable digital economy. The critiques against it, though real, are not insurmountable. History shows that when labor is recognized, valued, and organized, transformative change follows.

Now is our moment to realign the social contract for the AI era. Just as the factory floor forged new protections for physical labor, the digital arena must forge dignity for data labor.

Let us build the scaffolding now—so the future we inherit is not one of passive surveillance, but one of active, dignified participation.

KEY TAKEAWAYS

- Universal Basic Income (UBI) provides a financial safety net that reduces poverty, improves well-being, and allows people to navigate economic transitions with dignity.
- Data dividends complement UBI by recognizing users as contributors to the AI economy and compensating them for the data they already generate.
- Together, UBI and data dividends represent a shift from dependency to participation—combining unconditional support with earned agency.
- The concept of participatory compensation is not new; examples like royalties, loyalty programs, and crowdsourced innovation platforms show it can work.
- Although critics raise valid concerns, advances in technology, clear economic precedents, and historical analogs suggest a data dividend economy is feasible and necessary.
- Just as the industrial revolution gave rise to new labor protections, the digital age calls for new institutions that ensure fairness, dignity, and economic inclusion.

LOOKING AHEAD

In the next chapter, we'll explore why this proposal isn't just ethically sound, but also strategically aligned with the interests of tech companies. We'll look at why platforms that embrace data dignity—through consent, compensation, and transparency—may not just survive, but thrive in the emerging digital economy.

CHAPTER 10
WHY I BELIEVE TECH WILL GO ALONG

WHEN TRUST BECAME THE PRODUCT

IN EARLY 2025, *the privacy-focused browser* **DuckDuckGo** *surpassed 100 million active users. Not because it had the flashiest interface or deepest pockets—but because it promised not to track, sell, or manipulate its users. What once seemed like a fringe value—privacy—had become a mainstream demand. Tech giants took notice. Shareholder reports began to cite "user trust" not just as a public relations asset, but as a growth metric.*

In an era of rising digital skepticism, tech companies have learned a basic truth: **people will walk away from platforms they no longer trust***. And without users, there is no data. Without data, there is no AI.*

Without users, there is no data. Without data, there is no AI.

∽

THE CASE FOR COOPERATION

At first glance, a data dividend model might seem threatening to Big Tech. Why would companies give up a slice of the pie they've long consumed freely? The answer is simple: because the alternative is worse.

Historical Lessons in Ignoring Contributors

History offers clear lessons: when powerful institutions ignore the dignity of contributors, the cost often comes due. During the industrial revolution, unsafe working conditions and exploitative labor led to mass strikes and the birth of the labor movement. In the 1911 Triangle Shirtwaist Factory fire in New York, 146 workers died due to locked exits and inadequate safety—galvanizing national outrage and leading to major labor reforms [1].

Tech's Regulatory Backlash

In the tech world, backlash against unchecked data collection has already led to significant regulatory responses. After the 2018 Cambridge Analytica scandal, Facebook faced $5 billion in fines from the FTC and was forced to alter its data practices [2].

When Companies Lead to Avoid Regulation

Moreover, when regulation lags, companies sometimes lead—not out of altruism, but necessity. In the wake of growing AI scrutiny, Amazon, Anthropic, Google, Inflection, Meta, Microsoft, and OpenAI pledged AI safety frameworks ahead of legislation, knowing public pressure and institutional trust were at stake [3].

User Disillusionment and Rising Risks

When displaced populations grow large and vocal—like gig workers protesting Uber policies, or creators rallying against AI art generators—the risk of severe legislation increases. In the EU, the 2023 AI Act imposes strict compliance burdens on companies deploying high-risk AI models, shaped in part by public concern and organized pressure.

From Free Gifts to Existential Threats

The writing is already there—etched by a growing tide of user frustration and broken trust. It began innocently enough: free email accounts, cloud storage, social media platforms that promised connection. At first, these services felt like gifts. But slowly, users came to understand the tradeoff—they weren't the customer; they were the product.

They began noticing the ads. Then the ads started following them. Email inboxes turned into marketing channels. YouTube learning sessions were broken by mid-roll interruptions. Browsing habits were tracked, sold, and resold across shadowy ad networks. Free tools became surveillance portals. And just when users realized how much value they were helping create, they were asked to pay to escape it—buy premium to block ads, pay extra for privacy, subscribe to platforms built on their own behavioral data.

And now, it's more than annoying. It's existential. The very data they generated—their clicks, comments, and corrections—has become the foundation of AI systems that are displacing their jobs. Writers see their tone replicated by machines. Coders watch their Stack Overflow answers fuel the models that threaten their roles. Call center workers are replaced by bots trained on decades of recorded interactions.

Rising Injustice and Implications for Companies

People who once thought they were merely being inconvenienced are waking up to the deeper injustice: the same platforms that profited from their presence are now automating them out of relevance.

This is not a future any company wants to inherit. Disempowered, disillusioned users become adversaries. They organize, legislate, and demand structural change. If companies don't proactively design participatory, fair systems now, they will face the same fate that 20th-century industrial giants did when they ignored the plight of factory labor: revolt, regulation, and reputational collapse. Consider General Motors in the 1970s, which faced intense labor unrest, strikes, and eventually damaging public scrutiny due to poor working conditions and a refusal to engage meaningfully with its workforce [4]. Or the fall of U.S. Steel, once an industrial titan, which failed to modernize and ignored worker demands for fair labor practices—leading to protracted strikes, lost market share, and eventual decline [5] [6]. These companies didn't collapse overnight, but their reputations eroded, their labor costs ballooned under adversarial conditions, and public trust waned. The parallels to today's digital platforms are not merely symbolic—they are cautionary blueprints.

Cooperation is not just a good idea. It's the only viable path forward.

Cooperation, by contrast, offers three major advantages—each rooted in self-interest as much as in ethics. If tech companies are to survive and thrive in the age of digital skepticism, they must choose participation over extraction.

1. **User Trust = Platform Longevity**

Trust isn't just a virtue—it's a survival strategy. When users feel deceived or exploited, they leave. We've seen it with platform declines like MySpace, whose inability to adapt to user needs and privacy expectations led to its rapid fall from prominence [7] [8]. Similarly, Facebook's younger user base has steadily eroded, driven in part by competition from newer, more engaging platforms like TikTok and Snapchat. According to Pew Research Center, teen usage of Facebook dropped from 71% in 2015 to just 32% in 2022, with many younger users citing the platform as outdated or irrelevant to their social habits [9]. Analysts have also noted that Facebook's association with invasive data practices and older demographics has contributed to its decline among youth [10]. But when platforms build trust through transparency and consent, they deepen engagement and loyalty.

Consider how Apple positioned itself as a privacy-first brand. In 2021, when it introduced App Tracking Transparency (ATT), advertisers were enraged—but users applauded. That bet paid off: Apple saw a surge in loyalty and differentiated itself from competitors increasingly under fire for data abuse. Trust, it turns out, is not just a soft benefit—it's a hard moat [11] [12] [13] [14] [15].

2. *Legal and Regulatory Shielding*

The regulatory landscape is tightening globally—from Europe's GDPR and Digital Services Act to California's CPRA. Complying is complex, expensive, and fraught with reputational risk. But platforms that embrace participatory models—where data use is transparent and consensual—are better insulated from scrutiny.

Rather than waiting for punitive enforcement, some companies have led proactively. Microsoft's adoption of GDPR-like policies globally—even for jurisdictions that didn't require it—positioned the company as a trusted enterprise provider. When firms show they can self-regulate responsibly, they buy regulatory goodwill and minimize compliance shocks.

In a world where regulators often lack the speed or expertise to police fast-moving innovation, companies that lead with user-first practices build their own buffer zones.

3. *Better Data, Better AI*

AI doesn't learn in a vacuum. It learns from us. And models trained on stale, biased, or manipulated data perform poorly—damaging brand reputation, user experience, and ROI.

But when users are invited into the loop—told how their data is used and given a stake in the system—they produce better inputs. Think of platforms like Duolingo or ReCAPTCHA, where user behavior simultaneously serves a training function. Now imagine that behavior incentivized—users correcting a chatbot's mistake, labeling an image, or offering nuanced feedback—because they know their contribution counts.

Participatory models don't just make data collection more ethical. They make the outputs of that data—AI systems, predictive models, personalization—more robust and inclusive. In a world where model quality determines market leadership, that's not just a benefit. It's a necessity.

Even Meta—often criticized for aggressive data practices—has taken a notable step toward openness with its release of the LLaMA series of large language models. Unlike most competitors that keep models proprietary, Meta has made LLaMA models open to the research community, allowing for broader scrutiny and collaboration. This move has been associated with faster community-driven innovation and model improvement, particularly in multilingual and low-resource contexts [16] [17]. While not purely altruistic, this kind of transparency can foster goodwill, build trust, and demonstrate that even large platforms can pivot toward more cooperative AI ecosystems.

EARLY SIGNS OF CHANGE

Some companies are already moving in this direction:

- **Apple** emphasizes on-device processing and privacy as a product feature. Aforementioned App Tracking Transparency (ATT) feature, launched in 2021, required apps to get user permission before tracking their data across other apps or websites. While it disrupted ad networks and drew criticism from digital advertisers, Apple's reputation among privacy-conscious users grew stronger, reinforcing its brand loyalty and sparking similar privacy-forward changes across the industry [18].
- **Mozilla** continues to advocate for ethical data stewardship through its Firefox browser, which includes Enhanced Tracking Protection by default and pushes forward innovations like Total Cookie Protection. Mozilla also invests in open-source and decentralized web initiatives, such as Mozilla Rally, a platform that allows users to donate browsing data to research projects in a privacy-preserving way—bridging personal agency with scientific benefit [19].
- **Brave** enables users to opt into privacy-respecting ads in exchange for Basic Attention Tokens (BAT), a blockchain-based token system. Users are compensated for their attention, and advertisers are granted access only to aggregated, anonymized data. Brave's model has faced early scalability challenges and limited advertiser adoption, but it offers a proof-of-concept for user-compensating ad ecosystems and has built a loyal, privacy-oriented user base [20].
- **OpenMined** is a community-driven organization developing tools for privacy-preserving machine learning. Through techniques like federated learning and differential privacy, OpenMined allows models to be trained without centralized access to raw user data. Though still largely in the research and prototype phase,

it demonstrates the feasibility of a decentralized, privacy-conscious AI future [21].
- **Ocean Protocol** builds decentralized data marketplaces, allowing users to publish, discover, and monetize datasets using blockchain and smart contracts. While adoption has been gradual, Ocean has partnered with institutions in sectors like healthcare and finance to explore real-world applications. Its tokenized model of data exchange highlights the potential for individuals and communities to gain a stake in the data economy [22].

These are early signals, not endgames. But they show a willingness to rethink extractive models—and point toward a future where consent and compensation are baked into the tech stack.

A BETTER RELATIONSHIP WITH USERS

Think of what came before: a model where users were tracked invisibly, their behavior sold in secret, and their value captured without consent. Platforms thrived not because they empowered users, but because users had few alternatives.

But that's changing. Trust is now a differentiator. Companies like Apple have made privacy a product feature—and won loyalty for it. Mozilla has built its brand on ethical stewardship. Meanwhile, platforms that pushed too far, like Facebook, have seen their reputations tarnished and their younger user bases erode.

Now imagine a new social contract: one where platforms disclose what data is collected, let users set permissions, and share in the resulting value. A model where data isn't just taken—but earned. A model where participation is the default—not exploitation.

This isn't science fiction. It's a design choice. And like all good design, it balances function, trust, and value.

If the 20th century was about industrial labor and the factory floor, the 21st will be about digital labor and the data stream. And just as companies that treated workers well outperformed those that didn't, the platforms that treat users as partners—not products—will be the ones that thrive.

KEY TAKEAWAYS

- Trust, transparency, and user agency are becoming strategic advantages—not just ethical aspirations—for tech companies.
- Participatory data models help improve user retention, data quality, and long-term business sustainability.
- History shows that ignoring contributor dignity leads to revolt, regulation, and decline—while cooperation builds resilience.
- Early adopters like Apple, Mozilla, Brave, and Meta (with its LLaMA model) show that benevolent design choices can enhance both reputation and innovation.

LOOKING AHEAD

In the next and final chapter, we'll move from theory to practice. What does it take to build real systems that embody data dignity? From consent frameworks to smart contracts and personal data vaults, we'll explore the technical, institutional, and cultural building blocks that make a fairer digital economy possible—starting now, and pointing the way forward.

CHAPTER 11
THE PATH TO DATA LIBERATION

THE DAY EVERYTHING CHANGED

IN APRIL 2025, two major pieces of legislation—the EU Data Act and the Digital Services Act—entered into force, marking a watershed moment in data transparency and user rights across Europe. These laws granted citizens new powers to access, understand, and control how their data is used, especially in public digital services and connected devices.

Under the EU Data Act, users of smart devices—from thermostats to mobility cards—gained the right to see what data was being collected and to direct how and with whom it could be shared. Meanwhile, the Digital Services Act required online platforms to clearly explain how algorithms and data use shaped user experiences.

The effect was immediate. Across the continent, digital service providers began building tools to meet these requirements, and citizens gained unprecedented visibility into data practices. The idea that people should have agency over their digital footprint was no longer radical. It was the law.

For years, privacy advocates had warned about the risks of data *exploitation. Now, with AI systems displacing workers and monetizing human behavior, a new coalition was forming—not just to regulate, but to rebuild the foundation of digital life.*

The idea that people should have agency over their digital footprint was no longer radical. It was the law.

∽

WHAT NEEDS TO HAPPEN

Building a fair data economy won't happen overnight. But if this book has shown anything, it's that change is both necessary and possible. Each chapter has explored how data fuels our economy, how users are being left out of its rewards, and how emerging tools and ideas can reverse that imbalance. What follows is not a wish list —but a set of logical, actionable outcomes that stem directly from the ideas we've explored together.

There are concrete, immediate steps that individuals, companies, and policymakers can take today:

1. **Recognize Data as Labor**

Everyday digital activity—searches, clicks, corrections, preferences—fuels AI development and corporate profit. This labor is currently uncompensated and invisible. Recognizing it formally as labor sets the foundation for fair value exchange.

Who must lead: Policymakers can initiate legal recognition, while companies and researchers can adjust internal models of value attribution. Civil society must continue to press the case in public discourse.

2. *Invest in Infrastructure for Data Attribution*

Value cannot be shared if it cannot be traced. We need systems that can log, permission, and credit digital contributions securely and scalably.

Examples include:

- Confidential computing for privacy-preserving analytics
- Personal data vaults for individual control
- Open standards for consent, attribution, compensation, and interoperability across platforms

Who must lead: Companies and open-source communities should build and adopt these tools. Governments can fund public infrastructure and enforce interoperability standards.

3. *Create Collective Mechanisms*

Just as workers organized into unions to negotiate better conditions, data contributors need collective structures. Data cooperatives, unions, and trusts can aggregate power, advocate for rights, and negotiate fair terms with platforms.

Who must lead: Civil society organizations, labor advocates, and forward-thinking platforms can incubate these models. Users can join or help form them.

4. *Support Ethical Tech Innovation*

The extractive model of "surveil and sell" isn't the only option. Innovations like zero-knowledge proofs, edge AI, and privacy-first design can uphold user agency and still power useful systems. Companies that embed dignity in their technology will have a competitive edge.

Who must lead: Policymakers can provide funding and favorable regulation. Startups and investors should prioritize business models that respect users. Technologists must center ethics in design.

5. *Push for Rights-Based Policy*

We need legislation that enshrines rights to understand, control, and benefit from personal data. These rights should be clear, enforceable, and universal—like the right to consent, delete, port, or be compensated for data use.

Who must lead: Lawmakers must draft and pass such laws, inspired by models like the EU's GDPR and Data Act. But they shouldn't act in isolation. To ensure these policies are both effective and implementable, they should collaborate closely with industry experts, researchers, and civil society—those with firsthand knowledge of emerging technologies and platform realities. Voters and advocacy groups must also hold them accountable to ensure public interest remains at the center.

WHAT YOU CAN DO NOW

This may feel like a systems problem—but users have more power than they think. Here are small but meaningful actions you can take right now:

- **Switch to privacy-respecting browsers, email, and messaging providers**: Use browsers like Brave[1], Firefox[2], or DuckDuckGo[3] that block trackers by default and prioritize user privacy. For email, consider services like Proton Mail[4], Tutanota[5], or Skiff[6] that offer end-to-end encryption and do not harvest data for advertising. For messaging, consider apps like Signal[7] or Session[8].
- **Ask your employer about data practices**: If you use workplace software (like Zoom, Slack, or Microsoft

Teams), ask HR or IT how employee data is handled. Are conversations recorded? Is behavior monitored? Request transparency and push for policies that respect staff privacy.
- **Join or support digital rights organizations**: Groups like Electronic Frontier Foundation (EFF)[9], Access Now[10], and MyData Global[11] are leading the fight for user agency. Donate, attend events, or share their work.
- **Talk to a friend or colleague**: Raise awareness by sharing what you've learned. Recommend a chapter or this book. Conversation builds momentum—and momentum drives change.
- **Vote with intention**: Research where candidates stand on digital rights. Look for support of data privacy, fair AI regulation, and user agency. Tools like Vote Smart[12] can help.
- **Opt out or speak up**: Use browser extensions like Privacy Badger[13] or uBlock Origin[14] to block trackers. If a platform doesn't offer meaningful privacy controls or compensation, let them know—and consider alternatives that do.

Every movement starts somewhere. This one starts with you.

A NEW BEGINNING

The age of invisible labor must end. For too long, we've contributed the raw material of the AI revolution—our clicks, voices, faces, habits—without recognition, reward, or recourse.

But this isn't just about reclaiming what was taken. It's about building something better.

A world where participation is meaningful. Where value is shared. Where dignity is restored.

This is the conclusion of our journey, but it's also the beginning of a new one. You now understand the mechanisms, the stakes, and the possibilities. And most importantly, you know your role in it.

The path to data liberation begins with a single truth:

You are not the product. You are the source.

∽

EPILOGUE

A LETTER FROM THE NEAR FUTURE

Maria Alvarez no longer scrubs hotel bathrooms for a living. That chapter of her life ended with the arrival of autonomous cleaning bots and productivity dashboards. But her story didn't end with job loss. It marked the beginning of something more powerful: a shift from invisibility to agency.

When a community pilot offered residents the chance to join a local data trust, Maria hesitated. She wasn't a tech person. She didn't see how her routines—her walking routes, her streaming preferences, her idle voice searches—held any value. But the community advocate framed it differently:

"You've already trained the machines. This is about getting credit for it. This is about building a future where people like you help make the rules."

With a few taps, Maria connected her personal data vault: browser history, energy usage, mobility data, and yes—even the steps she logged pushing her cleaning cart. The vault encrypted everything, anonymized her identity, and allowed her to choose which organizations could access it—and under what terms.

Soon, small but steady monthly deposits arrived. Not from wages. Not from government assistance. But from the value of her data—shared with her permission and protected by the data trust.

But the money wasn't the most powerful change. It was the shift in posture. Maria began attending digital town halls where local residents shaped the algorithms used in city services. She helped vote on data-sharing terms for public transit planning. She even contributed voice samples to a multilingual AI project—and was credited as a co-trainer.

She was no longer a passive user. She was a participant. A co-creator. A stakeholder in the systems that once ignored her.

Now, Maria helps train others. She mentors new participants, explains how smart contracts enforce fairness, and helps her community understand what it means to own a piece of the digital economy.

She doesn't just earn a living. She helps shape the rules.

Her story is fiction.

For now.

But everything in it exists in fragments—in research labs, local cooperatives, pilot policies, and emerging technologies. What's missing is connection: the public imagination to demand more, the civic will to align incentives, and the courage to move from critique to construction.

That's why this book was written.

So that when your daughter asks why her generation has no say in the digital world she inherits—you'll have an answer.

Or better yet: a plan.

APPENDIX A: RESOURCES FOR DIGITAL EMPOWERMENT

TOOLS & TECHNOLOGIES

- **Solid (Inrupt)** – Decentralized data pods for user-owned data
- **Nextcloud** – Self-hosted cloud platform for privacy-first storage
- **GrapheneOS / CalyxOS** – Privacy-respecting Android operating systems
- **Brave / Firefox** – Browsers with strong tracking protections
- **Proton / Tutanota** – Encrypted email services
- **Signal / Session** – Encrypted messaging apps
- **Mina Protocol / Ocean Protocol** – Blockchain-based data economies

ORGANIZATIONS TO WATCH

- **MyData Global** – Advocates for human-centric data ownership
- **The Aapti Institute** – Research and policy on data stewardship
- **Mozilla Foundation** – Promotes internet health and digital rights
- **OpenMined** – Decentralized AI with privacy-preserving training
- **Confidential Computing Consortium** – Advancing trusted execution environments for secure data processing

PROJECTS & PILOTS

- **RBC's Virtual Clean Room** – Enterprise confidential computing for finance
- **Estonia's e-Governance Platform** – Transparent, citizen-controlled data infrastructure
- **Apple's On-Device AI** – Illustrates commercial push toward local processing

APPENDIX B: A READER'S DIGITAL RIGHTS CHECKLIST

BEFORE SIGNING UP FOR ANY APP, PLATFORM, OR DEVICE, ASK:

1. Do I understand what data is being collected?
2. Can I opt out or limit what's collected?
3. Is the data stored securely—on my device or in a cloud I trust?
4. Who profits from my data?
5. Can I withdraw my data at any time?
6. Is there a way for me to share in the value it creates?

If you can't answer yes to at least 4 of these—walk away.

APPENDIX C: FREQUENTLY ASKED QUESTIONS (FAQ)

Q: Isn't my data already compensated through free services?

A: No. While platforms offer free access to services like email and social media, they monetize your behavior many times over—often without transparency or fair sharing. You don't receive market-based compensation or agency over how your data is used, even though your interactions train algorithms and generate ongoing profit.

Q: Would data dividends be enough to live on?

A: In the short term, probably not. But over time—especially as attribution systems mature and more companies adopt participatory models—data dividends could grow to provide meaningful supplemental income. In high-value contexts, they could even become salary-replacing for some households.

Q: Won't this just create more surveillance?

A: It doesn't have to. Confidential computing, personal data vaults, and zero-knowledge proofs allow systems to verify and compute data without exposing its contents. These tools empower users to participate in the data economy without compromising their privacy.

Q: Why would companies agree to this?

A: Because it's in their long-term interest. Trust, legal compliance, data quality, and user loyalty are powerful motivators. As seen with GDPR and shifts in public sentiment, extractive models carry reputational and regulatory risk. Participatory frameworks help companies future-proof their business models.

Q: Will this be complicated for users to manage?

A: No. As with Face ID or Touch ID, complexity can be abstracted away. Once the right interfaces and permissions infrastructure are in place, users will be able to participate with minimal effort—while still retaining meaningful control.

Appendix D: Glossary of Key Terms

- **Confidential Computing**: A technology that processes encrypted data within secure enclaves, ensuring no one —not even the service provider—can access the data even when in use.
- **Data Dignity**: The principle that individuals have the right to agency, transparency, and economic participation in how their data is used.
- **Data Dividend**: A form of compensation to individuals whose data contributes value to digital systems, particularly in AI and analytics.
- **Data Economy:** The system of economic activity that arises from the collection, processing, and monetization of data as a key resource.
- **Data Vault**: A user-controlled, secure environment where personal data is stored, managed, and permissioned.
- **Digital Colonialism**: A term describing how digital platforms, primarily based in a few powerful countries, extract data and economic value from global users without equitable return or governance.

- **Digital Sharecropping**: A term describing how users generate valuable data/content on platforms they don't control and from which they don't profit.
- **Trusted Execution Environment (TEE)**: A secure area of a processor used in confidential computing to protect data during computation.
- **Universal Basic Income (UBI)**: A policy model that provides all citizens with a regular, unconditional sum of money, typically from public funds.

APPENDIX E: ENDNOTES

3. THE AUTOMATION AVALANCHE

1. **McKinsey Global Institute.** (2017, December). *Jobs lost, jobs gained: Workforce transitions in a time of automation* (Executive summary). McKinsey & Company. Retrieved from https://www.mckinsey.com/featured-insights/future-of-work/jobs-lost-jobs-gained-what-the-future-of-work-will-mean-for-jobs-skills-and-wages.
2. **Organisation for Economic Co-operation and Development (OECD).** (2021, January). *What happened to jobs at high risk of automation?*. Retrieved from https://www.oecd.org/officialdocuments/publicdisplaydocumentpdf/?cote=DELSA/ELSA/WD/SEM(2021)2&docLanguage=En.
3. **Goldman Sachs Research.** (2023, March 27). *The potentially large effects of artificial intelligence on economic growth* (Global Economics Analyst report). Goldman Sachs. Retrieved from https://www.goldmansachs.com/insights.
4. **Instacart.** (n.d.). *Don't let the crow guide your routes.* Instacart Technology Blog. Retrieved from https://tech.instacart.com/dont-let-the-crow-guide-your-routes-f24c96daedba.
5. **Katz, B.** (2021, November 30). *Scaling a routing algorithm using multithreading and ruin-and-recreate.* DoorDash Engineering Blog. Retrieved from https://careersatdoordash.com/blog/scaling-a-routing-algorithm-using-multithreading-and-ruin-and-recreate.
6. **Vincent, J.** (2019, April 25). *How Amazon automatically tracks and fires warehouse workers for 'productivity'.* The Verge. Retrieved from https://www.theverge.com/2019/4/25/18516004/amazon-warehouse-fulfillment-centers-productivity-firing-terminations.
7. **Netflix Technology Blog.** (2021). *Rapid event notification system at Netflix.* Retrieved from https://netflixtechblog.com/rapid-event-notification-system-at-netflix-6deb1d2b57d1.
8. **Spotify Engineering.** (2022). *User intents and satisfaction with slate recommendations.* Retrieved from https://research.atspotify.com/user-intents-and-satisfaction-with-slate-recommendations.
9. **Data & Society.** (2019, February). *Algorithmic management in the workplace* [Explainer]. Retrieved from https://datasociety.net/wp-content/uploads/2019/02/DS_Algorithmic_Management_Explainer.pdf.
10. **World Economic Forum.** (2020). *Towards a reskilling revolution.* World Economic Forum. Retrieved from https://www3.weforum.org/docs/WEF_Towards_a_Reskilling_Revolution.pdf.

4. THE PROFITS YOU NEVER SAW

1. **Meta Platforms, Inc.** (2024). *Fourth Quarter and Full Year 2023 Financial Results.* Retrieved from https://s21.q4cdn.com/399680738/files/doc_financials/2023/q4/Earnings-Presentation-Q4-2023.pdf.
2. **Zuboff, S.** (2019). *The age of surveillance capitalism: The fight for a human future at the new frontier of power.* New York, NY: PublicAffairs.
3. **Alphabet Inc.** (2025). *Alphabet announces fourth quarter and fiscal year 2024 results.* Retrieved from https://abc.xyz/investor/static/pdf/2024Q4_alphabet_earnings_release.pdf.
4. **Williams, M., & Murgia, M.** (2024, February 9). *OpenAI hits $2bn annualised revenue milestone.* Financial Times. Retrieved from https://www.ft.com/content/79c79c19-1a9f-4a5a-871c-1c208b5c093d.
5. **Apple Inc.** (2024). *Introducing Apple's on-device foundation models.* Apple Machine Learning Research. Retrieved from https://machinelearning.apple.com/research/introducing-apple-foundation-models.

6. INFRASTRUCTURE FOR DATA DIGNITY

1. **Center for Digital Health Innovation, UCSF, Fortanix, Intel, & Microsoft Azure.** (2020, October 7). *UCSF, Fortanix, Intel, and Microsoft Azure utilize privacy-preserving analytics to accelerate AI in health care.* UCSF News Center. https://www.ucsf.edu/news/2020/10/418736/ucsf-fortanix-intel-and-microsoft-azure-utilize-privacy-preserving-analytics.
2. **Fortanix. (n.d.).** *Fortanix unlocks the power of confidential computing.* Fortanix. Retrieved July 20, 2025, from https://www.fortanix.com/blog/fortanix-unlocks-the-power-of-confidential-computing.
3. **Microsoft. (2024, February 29).** *Enabling data clean rooms with confidential computing.* Azure Architecture Blog. https://techcommunity.microsoft.com/blog/azurearchitectureblog/enabling-data-clean-rooms-with-confidential-computing/4020538.
4. **Microsoft.** (2025, February 28). *Cleanroom and multi-party data analytics.* Microsoft Learn. https://learn.microsoft.com/en-us/azure/confidential-computing/multi-party-data.
5. **Royal Bank of Canada.** (n.d.). *Introducing Arxis: The future of data collaboration is here.* RBC. Retrieved July 20, 2025, from https://www.rbcroyalbank.com/dms/enterprise/sai-collision/arxis.html.
6. **Intel.** (n.d.). *Intel® Software Guard Extensions (Intel® SGX).* Intel. Retrieved July 20, 2025, from https://www.intel.com/content/www/us/en/products/docs/accelerator-engines/software-guard-extensions.html.
7. **Arm.** (n.d.). *TrustZone for Cortex-A.* Arm. Retrieved July 20, 2025, from https://www.arm.com/technologies/trustzone-for-cortex-a.
8. **Arm.** (n.d.). *TrustZone for Cortex-M.* Arm. Retrieved July 20, 2025, from https://www.arm.com/technologies/trustzone-for-cortex-m.

9. **Microsoft**. (n.d.). *Azure confidential computing*. Microsoft Azure. Retrieved July 20, 2025, from https://azure.microsoft.com/en-us/solutions/confidential-compute/.
10. **Apple**. (n.d.). *Apple Platform Security*. Apple Support. Retrieved July 20, 2025, from https://support.apple.com/guide/security/secure-enclave-sec59b0b31ff/web.
11. **Royal Bank of Canada**. (n.d.). *Introducing Arxis: The future of data collaboration is here*. RBC. Retrieved July 20, 2025, from https://www.rbcroyalbank.com/dms/enterprise/sai-collision/arxis.html.
12. **e-Estonia**. (n.d.). *E-services and registries*. Retrieved July 20, 2025, from https://e-estonia.com/solutions/e-governance/e-services-registries/.

7. YOUR DATA, YOUR RULES

1. **Inrupt**. (n.d.). *Inrupt*. Retrieved July 20, 2025, from https://inrupt.com/.
2. **MyData**. (n.d.). *MyData operators*. Retrieved July 20, 2025, from https://mydata.org/operators/.

8. DATA AS LABOR, TRACKED AND PAID

1. **Meta Platforms, Inc.** (2024). *Fourth Quarter and Full Year 2023 Financial Results*. Retrieved from https://s21.q4cdn.com/399680738/files/doc_financials/2023/q4/Earnings-Presentation-Q4-2023.pdf.
2. **Bellan, R.** (2022, May 15). *Uber Eats pilots autonomous delivery with Serve Robotics, Motional*. TechCrunch. https://techcrunch.com/2022/05/15/uber-eats-pilots-autonomous-delivery-with-serve-robotics-motional/.
3. **Lienert, P., & Sharma, S.** (2022, April 4). *Logistics giants hedge their bets in uncertain U.S. self-driving truck race*. Reuters. https://www.reuters.com/business/autos-transportation/logistics-giants-hedge-their-bets-uncertain-us-self-driving-truck-race-2022-04-04/.
4. **Spice AI**. (n.d.). *Spice AI*. Retrieved July 20, 2025, from https://www.spice.ai/.
5. **Republik**. (n.d.). *Republik*. Retrieved July 20, 2025, from https://www.republik.gg/.
6. **Swash**. (n.d.). *Swash*. Retrieved July 20, 2025, from https://swashapp.io.
7. **Streamr**. (n.d.). *Streamr*. Retrieved July 20, 2025, from https://streamr.network/.
8. **OpenMined**. (n.d.). *OpenMined*. Retrieved July 20, 2025, from https://openmined.org.
9. **AI Now Institute**. (n.d.). *AI Now Institute*. Retrieved July 20, 2025, from https://ainowinstitute.org/.
10. **DataUnion Foundation**. (n.d.). *DataUnion Foundation*. Retrieved July 20, 2025, from https://www.dataunion.app/.
11. **MyData Global**. (n.d.). *MyData Global*. Retrieved July 20, 2025, from https://mydata.org/.
12. **Fairwork**. (n.d.). *Fairwork*. Retrieved July 20, 2025, from https://fair.work/.
13. **Self-Employed Women's Association**. (n.d.). *SEWA*. Retrieved July 20, 2025, from https://www.sewa.org/.

9. UBI ALONE ISN'T ENOUGH — WHY DIGNITY BEATS DEPENDENCY

1. **Wellbeing Economy Alliance**. (n.d.). *Finland universal basic income pilot*. WEAll. Retrieved July 20, 2025, from https://weall.org/resource/finland-universal-basic-income-pilot.
2. **VTT Technical Research Centre of Finland**. (n.d.). *Data markets for sustainable cities*. VTT. Retrieved July 20, 2025, from https://cris.vtt.fi/en/projects/data-markets-for-sustainable-cities.
3. **Alaska Department of Revenue**. (n.d.). *Permanent Fund Dividend Division*. Retrieved July 20, 2025, from https://pfd.alaska.gov/.
4. **Wellbeing Economy Alliance**. (n.d.). *Finland universal basic income pilot*. WEAll. Retrieved July 20, 2025, from https://weall.org/resource/finland-universal-basic-income-pilot.
5. **Government of Ontario**. (n.d.). *Ontario basic income pilot*. Retrieved July 20, 2025, from https://www.ontario.ca/page/ontario-basic-income-pilot.
6. **Stockton Economic Empowerment Demonstration**. (n.d.). *Stockton Economic Empowerment Demonstration*. Retrieved July 20, 2025, from https://www.stocktondemonstration.org.
7. **Meta Platforms, Inc**. (2024). *Fourth Quarter and Full Year 2023 Financial Results*. Retrieved from https://s21.q4cdn.com/399680738/files/doc_financials/2023/q4/Earnings-Presentation-Q4-2023.pdf.
8. **Amazon.com, Inc**. (n.d.). *Investor relations*. Retrieved July 20, 2025, from https://ir.aboutamazon.com/overview/default.aspx.
9. **Alphabet Inc**. (n.d.). *Investor relations*. Retrieved July 20, 2025, from https://abc.xyz/investor/.
10. **Apple**. (2023, November 2). *Apple reports fourth quarter results*. Apple Newsroom. https://www.apple.com/newsroom/2023/11/apple-reports-fourth-quarter-results/.

10. WHY I BELIEVE TECH WILL GO ALONG

1. **Cornell University ILR School**. (n.d.). *Remembering the 1911 Triangle Factory Fire*. Retrieved July 20, 2025, from https://trianglefire.ilr.cornell.edu.
2. **Federal Trade Commission**. (2019, July 24). *FTC imposes $5 billion penalty and sweeping new privacy restrictions on Facebook*. https://www.ftc.gov/news-events/news/press-releases/2019/07/ftc-imposes-5-billion-penalty-sweeping-new-privacy-restrictions-facebook.
3. **Regulation of AI in the United States**. (2025, June 10). In *Wikipedia*. Retrieved July 20, 2025, from https://en.wikipedia.org/wiki/Regulation_of_AI_in_the_United_States.
4. **Minchin, T. J**. (2022). *"A gallant fight": The UAW and the 1970 General Motors strike*. International Review of Social History, 68(1), 41–73. https://doi.org/10.1017/S0020859022000293.

5. **Bureau of Labor Statistics**. (1986, January). *Structural change in steel. Monthly Labor Review*, 109(1), 35–40. https://www.bls.gov/opub/mlr/1986/01/art2full.pdf.
6. **Bureau of Labor Statistics**. (1987, January). *The steel industry in the 1980s. Monthly Labor Review*, 110(1), 32–40. https://www.bls.gov/opub/mlr/1987/01/art3full.pdf.
7. **Hern, A**. (2011, October 26). *Myspace: The rise and fall of the social networking site*. The Guardian. https://www.theguardian.com/technology/2011/oct/26/myspace-rebrand-collapse.
8. **Williams, S**. (2015, February 5). *I worked at Myspace and witnessed its fall — here's what it was like*. Business Insider via Hacker News. https://news.ycombinator.com/item?id=9156348.
9. **Pew Research Center**. (2022, August 10). *Teens, social media and technology 2022*. https://www.pewresearch.org/internet/2022/08/10/teens-social-media-and-technology-2022/.
10. **Heath, A**. (2021, October 25). *Facebook is worried about losing young people*. The Verge. https://www.theverge.com/22743744/facebook-teen-usage-decline-frances-haugen-leaks.
11. **Apple**. (2021, April). *A day in the life of your data*. https://www.apple.com/privacy/docs/A_Day_in_the_Life_of_Your_Data.pdf.
12. **TechJury**. (2023, February 6). *Customer loyalty statistics*. https://techjury.net/industry-analysis/customer-loyalty-statistics/.
13. **O'Brien, S. A**. (2022, April 25). *Apple's App Tracking Transparency explained: Why advertisers are worried*. Vox. https://www.vox.com/recode/23045136/apple-app-tracking-transparency-privacy-ads.
14. **Competition and Markets Authority**. (2022, June). *Appendix J: Apple's and Google's privacy changes (e.g., ATT, ITP, etc.)*[Report]. GOV.UK. https://assets.publishing.service.gov.uk/media/62a229c2d3bf7f036750b0d7/Appendix_J_-_Apple_s_and_Google_s_privacy_changes__eg_ATT__ITP_etc__-_FINAL_.pdf.
15. **Competition and Markets Authority**. (2022, March 11). *Apple's privacy changes and their impact on competition*. GOV.UK. https://assets.publishing.service.gov.uk/media/62277271d3bf7f158779fe39/Apple_11.3.22.pdf.
16. **Meta**. (2024, July 18). *Open source AI is the path forward*. Meta Newsroom. https://about.fb.com/news/2024/07/open-source-ai-is-the-path-forward/.
17. **Soni, S**. (2024, July 23). *Meta unveils biggest Llama 3 AI model touting language, math gains*. Reuters. https://www.reuters.com/technology/artificial-intelligence/meta-unveils-biggest-llama-3-ai-model-touting-language-math-gains-2024-07-23/.
18. **Apple**. (2021, April). *A day in the life of your data*. https://www.apple.com/privacy/docs/A_Day_in_the_Life_of_Your_Data.pdf.
19. **Mozilla**. (n.d.). *Mozilla Rally*. Retrieved July 20, 2025, from https://rally.mozilla.org/.
20. **Brave Software**. (n.d.). *Brave*. Retrieved July 20, 2025, from https://brave.com/.
21. **OpenMined**. (n.d.). *OpenMined*. Retrieved July 20, 2025, from https://www.openmined.org/.
22. **Ocean Protocol**. (n.d.). *Ocean Protocol*. Retrieved July 20, 2025, from https://oceanprotocol.com.

11. THE PATH TO DATA LIBERATION

1. **Brave Software**. (n.d.). *Brave*. Retrieved July 20, 2025, from https://brave.com/.
2. **Mozilla**. (n.d.). *Firefox*. Retrieved July 20, 2025, from https://www.mozilla.org/firefox/.
3. **DuckDuckGo**. (n.d.). *DuckDuckGo*. Retrieved July 20, 2025, from https://duckduckgo.com/.
4. **Proton**. (n.d.). *Proton*. Retrieved July 20, 2025, from https://proton.me/.
5. **Tutanota**. (n.d.). *Tutanota*. Retrieved July 20, 2025, from https://tutanota.com/.
6. **Skiff**. (n.d.). *Skiff*. Retrieved July 20, 2025, from https://skiff.com/.
7. **Signal Foundation**. (n.d.). *Signal*. Retrieved July 20, 2025, from https://signal.org/.
8. **Session**. (n.d.). *Session*. Retrieved July 20, 2025, from https://getsession.org/.
9. **Electronic Frontier Foundation**. (n.d.). *Electronic Frontier Foundation*. Retrieved July 20, 2025, from https://eff.org/.
10. **Access Now**. (n.d.). *Access Now*. Retrieved July 20, 2025, from https://accessnow.org/.
11. **MyData**. (n.d.). *MyData*. Retrieved July 20, 2025, from https://mydata.org/.
12. **Vote Smart**. (n.d.). *Vote Smart*. Retrieved July 20, 2025, from https://votesmart.org/.
13. **Electronic Frontier Foundation**. (n.d.). *Privacy Badger*. Retrieved July 20, 2025, from https://privacybadger.org/.
14. **uBlock Origin**. (n.d.). *uBlock Origin*. Retrieved July 20, 2025, from https://ublockorigin.com/.

APPENDIX F: FURTHER READING

1. **Zuboff, S**. (2019). *The age of surveillance capitalism: The fight for a human future at the new frontier of power.* PublicAffairs.
2. **Mazzucato, M**. (2021). *Mission economy: A moonshot guide to changing capitalism.* Harper Business.
3. **Doctorow, C., & Giblin, R**. (2022). *Chokepoint capitalism: How big tech and big content captured creative labor markets and how we'll win them back.* Beacon Press.
4. **Nissenbaum, H**. (2010). *Privacy in context: Technology, policy, and the integrity of social life.* Stanford University Press.Jaron Lanier, *Who Owns the Future?*
5. **Organisation for Economic Co-operation and Development**. (2019). *Data governance in the public sector.* OECD Publishing. https://www.oecd.org/gov/data-governance-in-the-public-sector-4af1c7b7-en.htm.
6. **Ellingrud, K., Sanghvi, S., Dandona, G. S., Madgavkar, A., Chui, M., White, O., & Hasebe, P**. (2023, July 26). *Generative AI and the future of work in America.* McKinsey Global Institute. https://www.mckinsey.com/mgi/our-

research/generative-ai-and-the-future-of-work-in-america.
7. **Intel White Papers** on Confidential Computing. https://www.intel.com.
8. **ARM Technical Reports** on TrustZone and Secure Environments. https://www.arm.com.
9. **Microsoft Learn** on Azure Confidential Computing. https://www.microsoft.com.

This curated list is a starting point for deeper exploration into digital rights, ethical technology, and the economic implications of AI.

Acknowledgments

Writing this book has made me realize, more than ever, that I stand on the shoulders of giants. My career has been shaped and enriched by the many companies, mentors, and colleagues who have believed in me, challenged me, and helped me grow.

I am especially grateful to **Atmel**, **Microchip**, and **Microsoft**, where I was formed professionally and given the platform to collaborate with some of the brightest minds and colleagues across these companies and the globe.

Towards the vision of data dignity, I owe deep gratitude to my collaborators and peers across numerous companies and engineering disciplines. Each conversation, each debate, and each technical breakthrough has brought this vision closer to reality. Here is just a partial list of the companies and organizations I have collaborated with – each working on different components of the technologies, services, and standards that together form the building blocks needed to make this vision possible. This shows just how complex and collaborative it will be to bring data dignity to fruition:

Arm, Arrow, Atmel, AWS, Brightsight, Confidential Computing Consortium (CCC), DigiCert, Eurotech, Fortanix, Giesecke+Devrient, GlobalPlatform, Global Semiconductor Alliance (GSA), GlobalSign, Google, Hewlett Packard, Infineon, Intel, IntrinsicID (now part of Synopsys), KeyFactor, Linaro, Microchip, Microsoft, Moxa, NxP, ProvenRun, Qualcomm, Renesas, Riscure, STMicroelectronics, Samsung, Scalys, SecEdge (formerly Sequitur Labs), TrustCB, UL.

Thank you for believing in building technologies that advance dignity, agency, and inclusion for everyone.

About the Author

Eustace Asanghanwa is a visionary, technologist and a leading voice in the movement for data dignity. With a career spanning engineering, privacy-enhancing technologies, and digital infrastructure, he has long championed **confidential computing**—a breakthrough that allows individuals to contribute to the digital economy without giving up control of their data.

His work is grounded in optimism about technology's potential to serve both business and society. Eustace believes that a privacy-respecting digital economy is not only possible but advantageous—for users and for companies alike. He champions solutions that align user trust with enterprise innovation, helping build systems that are secure, inclusive, and economically sustainable.

This book took shape during a pivotal moment in Eustace's professional journey—one that sharpened his resolve to build more equitable digital systems. After receiving the now all-too-common message—**"your position has been eliminated"**—Eustace channeled his energy into capturing a vision he had explored through blogs, talks, and industry collaboration. As someone directly affected by automation, he brings a rare combination of technical expertise and lived experience to the conversation on AI, labor, and fairness.

Far from leaving the tech world behind, Eustace is doubling down—committed to advancing technologies that balance profit with principle. He welcomes opportunities to work with builders, companies, and policymakers to design a data economy where users are seen, valued, and rewarded.

To learn more or connect, visit: https://linkedin.com/in/eustacea

www.ingramcontent.com/pod-product-compliance
Lightning Source LLC
Chambersburg PA
CBHW070639030426
42337CB00020B/4074